T0213240

# Communications in Computer and Information Science 684

*Commenced Publication in 2007*
Founding and Former Series Editors:
Alfredo Cuzzocrea, Dominik Ślęzak, and Xiaokang Yang

More information about this series at http://www.springer.com/series/7899

Boulbaba Ben Amor · Faten Chaieb
Faouzi Ghorbel (Eds.)

# Representations, Analysis and Recognition of Shape and Motion from Imaging Data

6th International Workshop, RFMI 2016
Sidi Bou Said Village, Tunisia, October 27–29, 2016
Revised Selected Papers

Springer

*Editors*
Boulbaba Ben Amor
IMT Lille Douai/CRIStAL
Villeneuve d'Ascq
France

Faten Chaieb
CRISTAL Laboratory/Institut National des
 Sciences Appliquées et de Technologie
Tunis
Tunisia

Faouzi Ghorbel
CRISTAL Laboratory/Ecole Nationale des
 Sciences de l'Informatique
Manouba
Tunisia

ISSN 1865-0929          ISSN 1865-0937 (electronic)
Communications in Computer and Information Science
ISBN 978-3-319-60653-8          ISBN 978-3-319-60654-5 (eBook)
DOI 10.1007/978-3-319-60654-5

Library of Congress Control Number: 2017944322

Printed on acid-free paper

This Springer imprint is published by Springer Nature
The registered company is Springer International Publishing AG
The registered company address is: Gewerbestrasse 11, 6330 Cham, Switzerland

# Preface

An object's *shape* and its *motion* are important cues in understanding imaging data. Their study is an emerging research area with a plethora of key applications in medical diagnosis, scene understanding (e.g., video surveillance), affective and behavioral computing, computer animation, HMI, robotics, and cultural heritage conservation, to cite a few. The external boundary of an object is available in various forms, they include landmark configurations, object contours (closed curves), 3D surfaces, and dense point clouds. Their analysis requires precise mathematical representations, rigorous methodologies to filter out undesirable transformations, and sophisticated computational solutions. The *motion* in turn, describes the *shape*'s evolution or animation over time. One can cite the examples of facial shape evolution with a person's age, the evolution of a cancer cell over weeks or months, or the evolution of the body's skeletal or silhouette data over time and their use to predict a person's activity. Despite the development of mathematically solid methodologies, the question is still open from both methodological and application perspectives. The challenges are even more acute with the new fields of application such as human brain understanding, body kinematic data analysis for rehabilitation and physical therapy, esthetic and reparation surgery, and 3D/4D microscopy. In addition, the rapid development of sophisticated medical imaging devices, cost-effective depth (Kinect-like) cameras, and new generations of microscopes push forth to develop suitable mathematical representations and design relevant geometric tools and set up computational solutions for advanced modeling, analysis, and understanding.

The International Workshop on Representations, Analysis and Recognition of Shape and Motion from Imaging Data (RFMI)[1] is one of the earliest events launched to study these questions. Its ultimate goal is to promote interactions among researchers from different disciplines, differential geometry, statistics, topology, computer vision and animation, and medical imaging to share their scientific results. Thus, the workshop keeps contributing to strengthen the relationship between many research areas that meet in facing similar challenges and using common tools.

The current edition (sixth) of the workshop received the endorsement of the prestigious International Association of Pattern Recognition (IAPR) and the current volume was accepted for publication in the CCIS series of Springer. The Program Committee of the workshop received 24 papers, among them nine were accepted as regular papers (i.e., 37.5% acceptance rate) and five were presented as short papers. In addition, the workshop's program included four outstanding tutorial-oriented invited talks. International speakers were invited – Dr. Anuj Srivastava (IAPR Fellow, IEEE Fellow, Professor at Florida State University in USA) presented "On Advances in the Role of Differential Geometry in Computer Vision and Pattern Recognition"; Dr. Xavier Pennec (Senior Research Scientist at INRIA in France) talked about "Riemannian and

---

[1] http://www.arts-pi.org.tn/rfmi2016/index.html.

Affine Structures for Statistics on Shapes and Deformations in Computational Anatomy"; Dr. Stefanos Zafeirou (Senior Lecturer at Imperial College London in UK) focused on "Building the First Large-Scale 3D Morphable Model of Faces"; and Prof. Mubarak Shah (Founding Director of the Center for Research in Computer Vision and Professor at University of Central Florida in USA) reviewed "Spatio-temporal Graphs for Object Segmentation and Human Pose Estimation in Videos." Around 60 researchers attended the workshop, among them 35 young PhD and Master students.

The workshop was organized in the authentic village of Sidi Bou Said, located north Tunis in Tunisia during October 27–29, 2016.

May 2017                                                                        Faouzi Ghorbel
                                                                    Boulbaba Ben Amor
                                                                         Faten Chaieb

# Organization

## Workshop Co-chairs

Boulbaba Ben Amor      IMT Lille Douai/CRIStAL (UMR CNRS 9189), France
Faten Chaieb      National School of Computer Science (ENSI), Tunisia

## Honorary Chair

Faouzi Ghorbel      National School of Computer Science (ENSI), Tunisia

## Steering Committee

Anuj Srivastava      Florida State University, USA
Bernadette Dorizzi      Telecom SudParis, France
Bulent Sankur      Bogazici University, Turkey
Basel Solaiman      Telecom Bretagne, France
Ezzedine Zagrouba      ENIT/RIADI-GDL, Tunisia
Imed Riadh Ferah      ISAMM/RIADI-GDL, Tunisia
Jean-Luc Dugelay      Eurecom, France
Jean-Philippe Vandeborre      IMT Lille Douai/CRIStAL, France
Liming Chen      Ecole Centrale de Lyon/LIRIS, France
Mohammed Bennamoun      University of Western Australia, Australia
Mohamed Hammami      Miracl/Sfax Unuiversity, Tunisia
Sylvie sevestre-ghalila      CEA-LinkLab, France

## Program Committee

Hazem Wannous      Lille 1 University/CRIStAL, France
Majdi Jribi      ENSI/CRIStAL, Tunisia
Stefano Berretti      University of Florence, Italy
Asma Ben Abdallah      FM Monastir, Tunisia
Amel Ben Azza      Sup'Com, Tunisia
Antony Fleury      IMT Lille Doaui, France
Alain Hillion      Telecom Bretagne, France
Abdelhak M. Zoubir      Technische Universität Darmstadt, Germany
Azza Ouled Zaid      ISI, Tunisia
Bernadette Dorizzi      Telecom SudParis, France
Marius Bilasco      Lille 1 University/CRIStAL, France
Baiqing Xia      Trinity College of Dublin, Ireland
Chokri Ben Amar      ENIS/REGIM, Tunisia
Chafik Samir      ISIT CNRS, France
Di Huang      Baihang University, China

# Contents

## 2D Shape Analysis

# 3D Shape Registration and Comparison

# Local Feature-Based 3D Canonical Form

Hela Haj Mohamed[1(✉)], Samir Belaid[1], and Wady Naanaa[2]

[1] FSM, University of Monastir, Monastir, Tunisia
hajmohamedhela@yahoo.fr, Samir.Belaid@fsm.rnu.tn
[2] ENIT, El-Manar University, Tunis, Tunisia
wady.naanaa@gmail.com

**Abstract.** In this paper, we present a novel approach to compute 3D canonical forms which is useful for non-rigid 3D shape retrieval. We resort to using the feature space to get a compact representation of points in a small-dimensional Euclidean space. Our aim is to improve the classical Multi-Dimensional Scaling MDS algorithm to avoid the super-quadratic computational complexity. To this end, we compute the canonical form of the local geodesic distance matrix between pairs of a small subset of vertices in local feature patches. To preserve local shape details, we drive the mesh deformation by the local weighted commute time. When used as a spatial relationship between local features, the invariant properties of the Biharmonic distance improve the final results.

We evaluate the performance of our method by using two different measures: the compactness measure and the Haussdorf distance.

**Keywords:** 3D canonical forms · Multidimensional scaling · 3D non-rigid shape · Biharmonic distance

## 1 Introduction

In the last few years, the recognition task of non-rigid 3D shapes, invariant to object's pose, becomes a significant challenge for modern shape retrieval methods. The review of algorithms for non-rigid 3D shape retrieval is mainly classified into algorithms employing local features, topological structures, isometric-invariant global geometric properties, direct shape matching, and canonical forms. In the last category, many authors proposed to transform each deformable model into a canonical form invariant to the pose. This proposal allows the rigid shape descriptors to be used in non-rigid shape retrieval.

The Multidimensional Scaling Method (MDS) is one of the most used methods for computing 3D canonical forms. However, the main challenge of the MDS based methods remains the construction of canonical forms with well-preserved features and with low time-complexity computation.

This paper introduces a novel canonical form based on multidimensional scaling (MDS) by considering local interest points. Our main idea amounts to partitioning the global MDS problem into a set of sub-problems and then to

© Springer International Publishing AG 2017
B. Ben Amor et al. (Eds.): RFMI 2016, CCIS 684, pp. 3–14, 2017.
DOI: 10.1007/978-3-319-60654-5_1

generating local 3D canonical forms of the original one. More specifically, feature points of the target model are first detected and then a set of local patches is generated. Thereafter, these sub-parts are transformed into their 3D canonical sub-forms by solving nonlinear minimization problems. Assuming this solution enables the reduction of the high computational cost of geodesic distance between each pair of vertices. Eventually, a spatial relationship constraint is needed between partitions to obtain the final 3D canonical form.

The rest of the paper is organized as follows: In Sect. 2, we briefly present MDS-based techniques. In Sect. 3, we establish the mathematical background of the proposed method. Then, in Sects. 4 and 5, we detail our method and report our experimental results.

## 2    MDS-Based Techniques

MDS is widely considered as an efficient tool to solve MDS problems. The basic idea of MDS-based algorithms is to map the dissimilarity measure between a pair of features, described in an initial feature space, into the distance between the corresponding pair of points in a small-dimensional Euclidean space. In fact, MDS maps each feature $Y_i, i = 1, ..., N$ to its corresponding point $X_i, i = 1, ..., N$ in a $m$-dimensional Euclidean space $R^m$ by minimizing a given stress function:

$$E_S(X) = \frac{\Sigma_{i=1}^{N}\Sigma_{j=i+1}^{N}\omega_{i,j}(d_F(Y_i, Y_j) - d_E(X_i, X_j))^2}{\Sigma_{i=1}^{N}\Sigma_{j=i+1}^{N}(d_F(Y_i, Y_j))^2} \tag{1}$$

where $d_F(Y_i, Y_j)$ is the dissimilarity between features $Y_i$ and $Y_j$, $d_E(X_i, X_j)$ is the Euclidean distance between points $X_i$ and $X_j$ in $R^m$, and $w_{i,j}$ is a weighting coefficient. The difference between $d_F(Y_i, Y_j)$ and $d_E(X_i, X_j)$ plays the role of the objective function, (also designated by *stress function*), $E_S(X)$ to be minimized. To solve this non-convex minimization problem, many algorithms have already been proposed. As standard optimization algorithm, we may mention Classical MDS [1], Least Squares MDS [2], Fast MDS [3], Non-metric MDS [4], etc.

In 3D domain, the first computed canonical form is performed by Elad and Kimmel [5]. The authors proposed using the least squares multidimensional scaling to generate a canonical form for a given 3D mesh. Wherein the dissimilarity measure between features is calculated by the geodesic distances. To improve the quality of canonical forms, Lian et al. [6] created a feature preserving canonical form by considering MDS embedding results as references and then naturally deform the original meshes against them. Nevertheless, this method is sensitive to topological errors, although it is quite robust against mesh segmentation results. In addition, the methods suffer from the high computational time due to the computation of geodesic distances between all pairs of vertices. To circumvent this difficulty, meshes are approximately simplified to 2000 vertices before the MDS embedding procedure is applied. Nevertheless, this solution can affect the quality of the mesh as well as its local features.

More recent methods are based on local feature [7,8]. For instance, Pickup et al. [8] have suggested a linear time complexity method for computing a canonical form. The authors maximized the distance between pairs of detected feature points whilst preserving the mesh's edge lengths. This is achieved by using Euclidean distances between pairs of a small subset of vertices. The same authors, proposed to perform the unbending on the skeleton of the mesh, and use this to drive the deformation of the mesh itself. They successfully saved computational time, and reduced distortion of local shape detail. However, this method is sensible to topological errors that may corrupt the mesh.

## 3    Mathematical Background

The Classical MDS, proposed by Elad and Kimmel [5] is based on a square and symmetric distance matrix $D_F$ defined by

$$D_F = \begin{bmatrix} d_F^2(Y_1, Y_1) & \cdots & d_F^2(Y_1, Y_N) \\ \cdot & \cdots & \cdot \\ \cdot & \cdots & \cdot \\ \cdot & \cdots & \cdot \\ d_F^2(Y_N, Y_1) & \cdots & d_F^2(Y_N, Y_N) \end{bmatrix} \quad (2)$$

where $d_F(Y_i, Y_j)$ is the geodesic distance between the pair $Y_i$ and $Y_j$, computed using the fast marching method [15] in the feature space. The inner product matrix (i.e., the Gram matrix) $G_E$ is calculated in the embedded Euclidean space by

$$G_E = -\frac{1}{2} J D_F J \quad (3)$$

where

$$J = \frac{1}{N} \mathbf{1}\mathbf{1}^T \quad (4)$$

where $\mathbf{1}$ denotes the $N$-one vector.

The Euclidean embedding of these distances is then computed using the eigen-decomposition of the Gram matrix $G_E$. So, the classical MDS minimizes the following energy function, instead of the stress function $E_S(X)$:

$$E_{S1}(X) = \|Q(\Lambda - \overline{\Lambda})Q^T\|^2 \quad (5)$$

where $\|\bullet\|$ denotes the Frobenieus norm of the squared matrix elements, $\Lambda_{m \times m} = diag(\lambda_1, \lambda_2, ..., \lambda_m)$ are eigenvalues of $G_E$ ordered so that $\lambda_1 \geq \lambda_2 \geq ...\lambda_k \geq 0$, $Q$ denotes the matrix having as columns the corresponding eigenvectors and $m$ is the dimension of the embedded Euclidean space.

The same authors [5] improved their results by suggesting a standard optimization algorithm. The Least Squares technique uses the SMACOF (Scaling by Maximizing a Convex Function) (Borg and Groenen [2]) algorithm to minimize the following stress function $E_S(X)$:

$$E_S(X) = \Sigma_{i=1}^N \Sigma_{j=i+1}^N \omega_{i,j}(d_F(Y_i, Y_j) - d_E(X_i, X_j))^2 \quad (6)$$

where $N$ is the number of vertices, the $w_{i,j}$'s are weighting coefficients, $d_F(Y_i, Y_j)$ is the geodesic distance between vertices $Y_i$ and $Y_j$ in the original mesh, and $d_E(X_i, X_j)$ is the Euclidean distance between vertices $X_i$ and $X_j$ of the resulting canonical mesh $X$. The algorithm iterates until $|S(X_i) - S(X_{i+1})|$ is less than a user-defined threshold $\epsilon$. This MDS algorithm has a complexity of $\mathcal{O}(N^2 \times$ NumOfIterations).

In the present paper, we adopt the Least Squares technique and the SMA-COF algorithm to compute the 3D canonical form. Our solution decreases the computational cost of geodesic distance computation, by using a set of local dissimilarity matrix.

## 4  Our Contribution

The main idea behind our method is to construct canonical forms that are invariant to the pose of 3D shapes. It is easy to see that the MDS embedding results naturally deform original deformable parts to normalised poses. For this reason, we resort to calculating local canonical form for each salient patch. We start by detecting the limbs of a given model using an automatic and unsupervised 3D salient point detector. The later task is based on the Auto Diffusion Function (ADF) proposed in [10]. The scalar function ADF is defined on the mesh surface as:

$$ADF_t = K(x, x, \frac{t}{\lambda_1}) = \Sigma_{i=0} e^{-t\frac{\lambda_t}{\lambda_1}} h_i^2(x) \tag{7}$$

where $\lambda$ and $h$ are eigenfunctions of Laplace-Beltrami operator (LBO). ADF function is controlled by a single parameter $t$ which can be interpreted as feature scale. The local maximum of the ADF is proved to be the natural feature points of the shape. A demonstration of the invariance of extracted points to non-rigid transformation, scaling, occultation and their insensitive to noise is given in [11]. Figure 1 illustrates examples of feature points extraction based on the ADF function. The parameter $t$ was fixed to 0.1.

The next step is to partition the 3D mesh into local regions using a Voronoi diagram, where seeds are the set of feature points. We then compute the canonical form of the mesh by calculating the value of $\delta_{i,j}$, for all vertices in the same patch. The value of $\delta_{i,j}$ is equal to the length of the shortest path connecting vertex $i$ to vertex $j$. This aims to minimising the computation cost of the geodesic distance and avoid exploring all the paths between every two points in the original space. Thus, the estimated complexity of this algorithm is $\mathcal{O}(km_k^2)$, where $k$ denotes the number of local patches and $m_k$ is the number of vertices in the $k^{\text{th}}$ region. In order to speed up the computation of geodesic distances, we use the heat method proposed by Crane et al. in [13], which gives an approximation of the geodesic distance by exploiting heat kernels. Figure 2 illustrates the results obtained by applying the fast marching algorithm and heat geodesic method. In this work, we used the Matlab source code available on the web site of the book (Bronstein et al. [14]) to calculate the canonical form. Clearly, the heat method gives similar results of the embedding form and has high convergence speed.

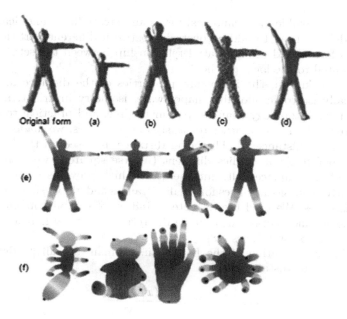

**Fig. 1.** Robustess for ADF detector to scale (a), occultation (b), noise (c), smooth (d), non-rigid deformations (e) and its reliance on shape (f).

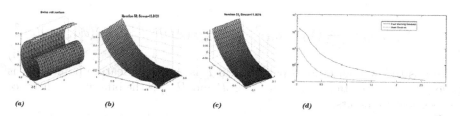

**Fig. 2.** Swiss roll surface (a), and its 3D canonical form using the classical MDS, (b) geodesic distance is computing by the fast marching algorithm and heat geodesic method(c). Their convergence speed is plotted is (d).

To preserve local features of the mesh, we do not treat all pairs of features in the same manner. However, we enforce a target weight between all connected vertices $i$ and $j$ in the same region. The value of $w_{i,j}$ is set according to the value of the commute-time distance $c_{i,j}$. Notes that this is the expected time taken by the random walk to travel from $i$ to $j$ in both directions. Therefore, if $c_{i,j}$ is small, then $\delta_{i,j}$ should also be small enough to minimize the stress function given in Eq. (1). The commute-time distance is defined as:

$$dc(i,j)^2 = \Sigma_{i=1}\frac{1}{\lambda_i}(h_i(x) - h_i(y))^2,\tag{8}$$

In order to assemble local canonical forms and create final canonical form of a given model, we need to add a constraint between different partitions. This constraint is formulated as the spatial relationship between the set of features points associated to the local patches.

As it is well known, the important properties of the distance are metric, smooth, locally isotropic, globally shape-aware, isometry invariant, insensitive to noise and small topology changes, parameter-free, and practical to compute on a discrete mesh. Thus, in order to satisfy our exigences, we made the choice of the biharmonic distance [9]. The latter distance, is based on the biharmonic differential operator, but applies different (inverse squared) weighting to the eigenvalues of the Laplace-Beltrami operator, which provides a nice trade-off between nearly geodesic distances for small distances and global shape-awareness for large distances. We used the discrete definition of the biharmonic distance based on the common cotangent formula discretization of the Laplace-Beltrami differential operator on meshes [18].

Given the discretization of the Laplacian, Lipman and all [9] defined the discrete Biharmonic distance as:

$$dB(x,y)^2 = \Sigma_{i=1} \frac{(h_i(x) - h_i(y))^2}{\lambda_i^2} \tag{9}$$

roughly speaking, we reformulate the final stress function to be minimised as:

$$S(X) = \Sigma_k \Sigma_{i,j \in P_k} \frac{c_{i,j}^2}{\delta i, j^2} (\delta_{ij} - d_{i,j})^2 + \frac{1}{2} \Sigma_{i,j \in F} (ADF_i + ADF_j)(dB_{i,j} - d_{i,j})^2 \tag{10}$$

where $k$ is the number of local patches and $F$ is the set of feature points. $\delta_{i,j}$ is the local geodesic distances in each region $P$ and $dB_{i,j}$ is the biharmonic distance between feature points.

We can ensure that our solution gives good results based on the comparison present in [10]. Thus, geodesic distance admits desirable local properties and gradually increases in the neighborhood of the source vertex (see Fig. 3). However, it is not globally shape-aware. In addition, the calculation of the biharmonic distances between extremities of the object is much faster than geodesic distances.

Figure 4 shows an overview of our algorithm (Fig. 4e) of a given 3D object (Fig. 4a). The scalar function ADF is used as an unsupervised detector of feature points (Fig. 4b and c). In Fig. 4c, we represent the overall geodesic distance computing from the local patches. For $n = 9501$ vertex, we use 17 million values in the dissimilarity matrix instead of $n^2$, i.e. approximately a reduction of 80%. Examples of 3D canonical forms produced by our method are also given in Fig. 5 for different 3D models.

The computational complexity of the 3D canonical form methods depend on the time complexity of distances calculation (matrices $\delta$ and $dB$ in Eq. (10) and the SMACOF algorithm (see Sect. 3).

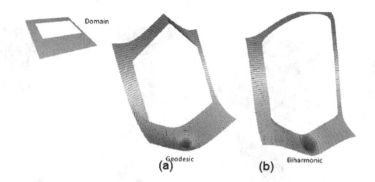

**Fig. 3.** Distances measured on an Euclidean domain (top-left). We visualize the distance field from a single source point to all other points as a height function over this Euclidean domain; (a) Geodesic distance, and (b)biharmonic distance. The geodesic distance possesses non-smooth curve of points and the isolines are not "shape-aware" far from the source. Biharmonic distance balances the "local" and "global" properties [10].

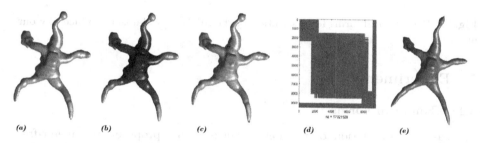

**Fig. 4.** Main steps of our procedure that employs the local features (c) extracted by the ADF scalar function defined on the mesh surface (b) to generate final 3D canonical form of the original model (a).

Geodesic based methods calculate the geodesic distance between all pairs of vertices using the fast marching algorithm which has a time complexity of $\mathcal{O}(N^2 \log N)$, where $N$ is the total number of mesh vertices. Our method calculates geodesic distances between only points in the same local patch rather than all pairs. For a local patch on $n$ vertices on average, with $n \ll N$, the time complexity is $\mathcal{O}(n^2 \log n)$. So, the total computational complexity of distance matrices has $\mathcal{O}(mn^2 \log n + m^2)$, where $m$ is the number of local patches. Note that $m$ is very small compared to $N$, (for instance, we have $m = 5$ and $N = 60000$, for the human body shape).

In addition, the SMACOF algorithm has a computational complexity of $\mathcal{O}(N^2)$, when the distance between all pairs of points is used. Our Algorithm lowers this complexity to $\mathcal{O}(mn^2)$.

**Fig. 5.** 3D Canonical forms for a selection of the SHREC'15 dataset produced by our method.

## 5      Experiments

### 5.1      Canonical Forms

We base the evaluation of the performance of the proposed parameter-free method for 3D canonical forms computation on the SHREC'15 Track [17]. The later is a new benchmark for testing algorithms at creating canonical forms for use in non-rigid 3D shape retrieval.

The dataset contains models from both the SHREC11 non-rigid benchmark [19] and the SHREC14 non-rigid humans benchmark [20]. The total number of meshes in the dataset is 100 meshes, split into 10 different shape classes. Each shape class contains a mesh in 10 different non-rigid poses.

Authors attempted to quantify how much the canonical forms distort the mesh's original shape, and the consistency of the different canonical forms of the same shape in different poses. They proposed two measures:

– The compactness measure is calculated as $\frac{V^2}{A^3}$, where V is the mesh volume and A is the mesh surface area. The difference in the compactness measure between the original mesh and the canonical form quantify the amount of distortion of the original shape.
– The Hausdorff distance is computed using iterative closest point matching to align each pair of shapes of the same class. This measure quantify the consistency of the canonical forms across models of the same shape.

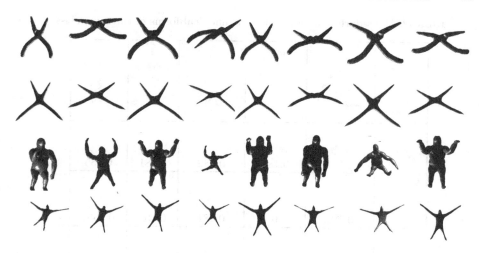

**Fig. 6.** Example of canonical forms of each mesh.

**Table 1.** Error in compactness between the canonical forms and the original meshes for each class. The values have been multiplied by $10^4$.

| Class | Mean Error | Standard deviation |
|---|---|---|
| Santa | 1.994 | 0.633 |
| Horse | 4.344 | 1.360 |
| Dog1 | 1.907 | 0.309 |
| Bird2 | 0.586 | 0.672 |
| Laptop | 0.336 | 0.741 |
| Female1 | 5.726 | 1.721 |
| Female2 | 5.281 | 1.390 |
| Male1 | 4.551 | 1.076 |
| Child body | 3.963 | 1.441 |
| Male2 | 4.246 | 1.445 |
| Whole | 3.3246 | 2.102 |

Several canonical forms of two objects with non-rigid deformations are shown in Fig. 6. Our method successfully producing canonical forms of each shape by eliminating non-rigid deformations and stretched out their extremities. The obtained results aim to standardize its pose.

Table 1 shows the compactness measure of our method for each class in the dataset at preserving the shape of the original model. Our method performs better at preserving compactness of all the dataset. This highlights the advantage of using local features instead of global features.

**Table 2.** Haussdorff distances between canonical forms of the same class.

| | | | | | |
|---|---|---|---|---|---|
| | 0 | 0.181 | 0.182 | 0.181 | 0.165 |
| | 0.181 | 0 | 0.181 | 0.181 | 0.1657 |
| | 0.182 | 0.181 | 0 | 0.1658 | 0.1657 |
| | 0.181 | 0.181 | 0.1658 | 0 | 0.1658 |
| | 0.165 | 0.1657 | 0.1657 | 0.1658 | 0 |

**Table 3.** Error in compactness between the canonical forms and the original meshes for each method. The values have been multiplied by $10^4$.

| | Classical MDS | Fast MDS | Least-squares MDS | Non-metric MDS | Skeletons | Euclidean random | Euclidean normalised | 1-Form MDS | Our method |
|---|---|---|---|---|---|---|---|---|---|
| Mean Error | 8.44 | 6.60 | 4.77 | 5.17 | 0.71 | 1.86 | 2.97 | 4.89 | 3.32 |
| Standard deviation | 4.420 | 3.384 | 3.595 | 3.754 | 0.907 | 1.362 | 1.231 | 3.712 | 2.102 |

Furthermore, we used the Meshlab [21] to compute the Hausdorff distance between a few of models of the same shape class of the SHREC'15 dataset. Table 2 shows small distances between the associated canonical forms, despite the presence of non-rigid deformations. These results promote the use of canonical forms in the non-rigid object recognition task.

In addition, we compare our results against those submitted to the SHREC'15 canonical forms benchmark [17]. Table 3 shows that our local method outperforms geodesic based methods and cause less shape distortion. However, our method is competitive with methods based on Euclidean distance. Yet, we only use local features at the ends of the shape members. In addition, the parameters of our method are fully automatic and free.

## 6  Conclusion

In this paper, we presented a novel method to construct canonical forms for 3D models based on the classical MDS method. Our contribution is to divide the problem of 3D canonical form embedding into sub-problems. To solve the resulting optimisation sub-problems, we added a spatial constraint between local features. We took advantage of the good properties of the Biharmonic distance to add relationship between sub-problems. In addition, we proposed a dynamic setting of the weight values in the stress function, according to the values of the dissimilarity matrix and the commute-time weight.

Experiments demonstrated the efficiency of our algorithm to construct an invariant 3D canonical form for a 3D surface. Our method preserve local and global features between original surfaces and MDS embedded surfaces and is also able to achieve the same bending invariant pose as the previous state-of-the-art. Yet, it involves far less shape distortion than other geodesic based methods.

As future works, we propose to test the performance of our method in the retrieval task on a recent canonical forms benchmark.

## References

1. Cox, M.A., Cox, T.F.: Multidimensional Scaling. Chapman and Hall, London/ New York (1994)
2. Borg, I., Groenen, P.: Modern Multidimensional Scaling: Theory and Applications. Springer, Heidelberg (1997)
3. Faloutsos, C., Lin, K.D.: A fast algorithm for indexing, data mining and visualisation of traditional and multimedia datasets. In: Proceedings of ACM SIGMOD, pp. 163–174 (1995)
4. Katz, S., Leifman, G., Tal, A.: Mesh segmentation using feature point and core extraction. Vis. Comput. **21**, 8–10 (2005)
5. Elad, A., Kimmel, R.: On bending invariant signatures for surface. IEEE Trans. Pattern Anal. Mach. Intell. **25**(10), 1285–1295 (2003)
6. Lian, Z., Godil, A., Xiao, J.: Feature-preserved 3d canonical form. Int. J. Comput. Vis. **102**, 221–238 (2013)
7. Wang, X.-L., Zha, H.: Contour canonical form: an efficient intrinsic embedding approach to matching non-rigid 3D objects. In: Proceedings of the 2nd ACM International Conference on Multimedia Retrieval, Article No. 31 (2012)
8. Pickup, D., Sun, X., Rosin, P.L., Martin, R.R.: Euclidean-distance-based canonical forms for non-rigid 3D shape retrieval. Pattern Recognit. **48**(8), 2500–2512 (2015)
9. Lipman, Y., Rustamov, R.M., Funkhouser, T.A.: Biharmonic distance. ACM Trans. Graph. (TOG) **29**(3), 27 (2010)
10. Gębal, K., Bærentzen, J.A., Aanæs, H., Larsen, R.: Shape analysis using the auto diffusion function. Comput. Graph. Forum **28**(5), 1405–1413 (2009)
11. Mohamed, H.H., Belaid, S.: Algorithm BOSS (Bag-of-Salient local Spectrums) for non-rigid and partial 3D object retrieval. Neurocomputing **168**, 790–798 (2015)

12. Papadimitriou, C.H.: An algorithm for shortest-path motion in three dimensions. IPL **20**, 259–263 (1985)
13. Crane, K., Weischedel, C., Wardetzky, M.: Geodesics in heat: a new approach to computing distance based on heat flow. ACM Trans. Graph. 32(5), Article No. 152 (2013)
14. Bronstein, A.M., Bronstein, M.M., Kimmel, R.: Numerical Geometry of Non-Rigid Shapes. Monographs in Computer Science. Springer, Heidelberg (2009)
15. Kimmel, R., Sethian, J.A.: Computing geodesic paths on manifolds. Proc. Natl. Acad. Sci. **95**, 8431–8435 (1998)
16. Surazhsky, V., Surazhsky, T., Kirsanov, D., Gortler, S.J., Hoppe, H.: Fast exact and approximate geodesics on meshes. ACM Trans. Graph. **24**(3), 553–560 (2005)
17. Pickup, D., et al.: SHREC 15 track: canonical forms for non-rigid 3D shape retrieval (2015)
18. Grinspun, E., Gingold, Y., Reisman, J., Zorin, D.: Computing discrete shape operators on general meshes. Eurograph. Comput. Graph. Forum **25**(3), 547–556 (2006)
19. Lian, Z., Godil, A., Bustos, B., Daoudi, M., Hermans, J., Kawamura, S., Kurita, Y., Lavoue, G., Nguyen, H.V., Ohbuchi, R., Ohkita, Y., Ohishi, Y., Porikli, F., Reuter, M., Sipiran, I., Smeets, D., Suetens, P., Tabia, H., Vandermeulen, D.: SHREC 11 track: shape retrieval on non-rigid 3D watertight meshes. In: Proceedings of the 4th Eurographics Conference on 3D Object Retrieval, pp. 79–88 (2011)
20. Cheng, Z., Lian, Z., Aono, M., Hamza, A.B., Bronstein, A., Bronstein, M., Bu, S., Castellani, U., Cheng, S., Garro, V., Giachetti, A., Godil, A., Han, J., Johan, H., Lai, L., Li, B., Li, C., Li, H., Litman, R., Liu, X., Liu, Z., Lu, Y., Tatsuma, A., Ye, J.: Shape retrieval of non-rigid 3D human models. In: Proceedings of the 7th Eurographics Workshop on 3D Object Retrieval, pp. 101–110 (2014)
21. Meshlabv1.3.0 (2011). http://meshlab.sourceforge.net

# Accurate 3D Shape Correspondence by a Local Description Darcyan Principal Curvature Fields

Ilhem Sboui[✉], Majdi Jribi[✉], and Faouzi Ghorbel[✉]

CRISTAL Laboratory, GRIFT Research Group,
National School of Computer Sciences, la Manouba,
University, Manouba, Tunisia
ilhem.sboui@gmail.com, {majdi.jribi,faouzi.ghorbel}@ensi.rnu.tn

**Abstract.** In this paper, we propose a novel approach for finding correspondence between three-dimensional shapes undergoing non-rigid transformations. Our proposal is based on the computation of the mean of curvature fields values on a local parametrization constructed around interest points on the surface. This local parametrization corresponds to the Darcyan coordinates system. Thereafter, correspondence is found by measuring the $L_2$ distance between obtained descriptors. We conduct the experimentation on the full objects of the Tosca database which contains a set of 3D objects with non-rigid deformations. The obtained results show the performance of the proposed approach.

**Keywords:** 3D shapes · Correspondence · Darcyan coordinates system · Principal curvatures

## 1 Introduction

Finding correspondence among three-dimensional shapes undergoing non-rigid transformations is one of the most fundamental recent problems in computer vision. It is actually a highly active research area, since it represents a key task in diverse applications such as motion tracking and recognition, shape interpolation and morphing, space-time reconstruction, shape retrieval and videos indexing.

Formally, matching pairs of shapes consists on establishing a map $f : S \to T$ between two given surfaces $S$ and $T$ which are semantically equivalent or their geometrical properties are similar.

Finding a correspondence between shapes in the rigid transformations case has been efficiently treated. It consists on the estimation of a rotation and a translation. However, matching two shapes undergoing non-rigid deformations, still, remains a challenging problem. Unlike the rigid case, the non-rigid one involves an important number of freedom degrees increasing with the number of matched points and thus, estimating these non-rigid transformations is, always, a hard task.

© Springer International Publishing AG 2017
B. Ben Amor et al. (Eds.): RFMI 2016, CCIS 684, pp. 15–26, 2017.
DOI: 10.1007/978-3-319-60654-5_2

The process of non-rigid shapes matching could be formulated as an optimization problem. It consists on finding the full correspondence mapping all point of one surface to their equivalent points on a second one with the minimal distortion.

The non-rigid correspondence methods can be classified on two major categories. We consider full and partial correspondence. The full or the complete one aims mapping the entire surfaces. While Partial matching is more complex, since it requires identifying optimal sub-parts or regions on shapes for which a right correspondence can be found.

According to the matched points resolution, the matching task can be also distinguished by dense or sparse. For the dense correspondence, the goal is to establish mapping between a large number of points on surface or even to find point-to-point matching of the corresponding shape. The sparse one is defined for a small number of elements or a set of feature points locally described. Besides, considering smaller sets of points permits to reduce the computational complexity comparing with the dense correspondence. Note that these categories of shapes matching may be combined, hence full as well as partial correspondence can be whether sparse or dense.

This challenging task has been widely addressed in the literature. Van Kaick et al. [20] proposed a detailed survey on 3D shapes matching methods.

For the full correspondence category, notably the sparse resolution, diverse methods have been proposed to tackle this problematic. The most common ones of these methods consist first on extracting a set of feature points and then constructing intrinsic surface descriptors.

Within this context, Ovsjanikov et al. [13] proposed an approach which relies on matching feature points in a space of a heat kernel for a given point on a surface and then the correspondence is obtained by searching the most similar heat kernel maps. Otherwise, Funkhouser et al. [17] used a Mobius voting approach. This approach consists on applying a Mobius transform for triplets of points, then generating conformal maps and voting for each couple of correspondences. The pairs with high votes are hence matched. On the other hand, Zhang et al. [21] proposed a method based on searching for correspondences while minimizing alignment and deformation error. Other alternative for the sparse correspondence searching, is presented by Bronstein et al. [2], by introducing the generalized multidimensional scaling (GMDS) which allows finding the minimum-distortion embedding of one surface into another. In the same context, Sahillioğlu et al. [16] presented a method based on greedy optimization of an isometric distortion function.

For the dense resolution of full matching, various methods in the literature seek to find correspondences for all points on a surface. Kim et al. [8] proposed to combine multiple low-dimensional intrinsic maps to produce a blended map. They, then, associated confidence and consistency weights to each map and find the best blending to establish a final correspondence. For the same purpose, Bronstein et al. [3] replaced the geodesic distance in the GromovHausdorff framework by the diffusion distance. Jiang et al. [7] proposed to embed the shapes into a spectral domain, and, then, to find the correspondence using a non-rigid variant of the ICP (Iterative Closest Point) algorithm. Very recently, Lähner et al. [10] proposed an algorithm for non-rigid 2D-to-3D shape matching, which consists on finding the shortest circular path on the product 3-manifold of the surface and the curve. In [14], Sahillioğlu et al. implemented an optimization mechanism of the method proposed by Sahillioğlu et al. [16] idea improved within the EM (Expectation-Maximization) framework and coupled with a more sophisticated sampling scheme.

Furthermore, for the partial category, with either dense or sparse resolution, different approaches exist for searching the correspondent parts of two given shapes. The common strategy to establish partial correspondence, for some methods, is to represent regions by descriptors and, then, to search the similarity between them. In this context, Funkhouser et al. [5] introduced a priority-driven algorithm for searching similar shapes from a large database of 3D objects. The authors used a priority queue of potential sets of partial matches sorted by a cost function representing feature dissimilarity and geometric deformation. Van Kaick et al. [19] explored the bilateral map in order to present a local shape descriptor whose region of interest is defined by two feature points instead of one unlike classical descriptors. Van Kaick et al. proved that their approach is more effective for partial matching but it's not the case when the shapes have strong intrinsic symmetries.

Hence some other alternatives compute a partial correspondence without relying on shape descriptors, such as the method of Bronstein et al. [1] whose main contribution is a framework for regularized partial matching of shapes taking into account three criteria, which are the regularity, the similarity and the size of parts. The authors relied on the Mumford-Shah functional to formulate the regularization criterion. Moreover, Sahillioğlu et al. [15] proposed a rank-and-vote-and-combine (RAVAC) algorithm that identifies and ranks potentially correct matches by exploring the space of all possible partial maps between shape extremities.

We propose in this paper a novel method for the sparse correspondence between 3D shapes undergoing non-rigid transformations. Our approach consists on an intrinsic local description of surfaces extremeties based on the construction of local discrete representation known by Darcyan Coordinates System. For each discrete representation around an extreme point on the surfaces, principal curvatures field are computed as well. The most similar point descriptors are therefore matched in the mean of the $L_2$ distance.

Thus, the present paper is organized as follow: We present in the second section the proposed descriptor construction steps. The next section is consecrated to the representation of our 3D local matching approach. Experimentations on 3D objects from the Tosca database and results discussion are illustrated in the fourth section.

## 2    3D Local Surface Description

We propose to locally describe the partial part of a surface around an extreme point of the surface with curvature. Since two acquisitions of the same object can lead to two different meshes, it is necessary first of all to extract a local representation around the extreme point which is invariant under the initial parametrization of the mesh. We propose to use the Darcyan representation. We present in this part all the steps of the proposed approach construction.

### 2.1    Brief Recall of the Darcyan Representation

Here, we recall the construction process of the well known Darcyan Coordinates System introduced by D'Arcy Thompson [18]. The parametric surface representation based on these coordinates system relatively to a given reference point on surface is, in fact, obtained by the superposition of the geodesic level curves around the reference point and the radial lines coming from the same point.

Let $S$ be a two dimensional differential manifold, and let consider $U_r$ the geodesic potential field coming from a reference point $r$ on $S$. The function $U_r : S \rightarrow R^+$ computes for any point $p$ on $S$ the length of the geodesic curve joining it to the reference point $r$. This function is well defined, since a geodesic curve between two points of a 2D differential manifold exists [4].

**Construction of the Geodesic Level Curves.** A geodesic level curve of value equal to $\lambda$ around a reference point $r$ on the surface $S$ can be formulated as follows:

$$L_r^\lambda = \{p \in S; U_r(p) = \lambda\} \tag{1}$$

$L_r^\lambda$ is materialized by a set of all points on $S$ having the same geodesic distance $\lambda$ from $r$. Therefore, the surface S can be approximately reconstructed by all geodesic level curves, as shown in Fig. 1, so that, $S \approx \cup_\lambda L_r^\lambda$.

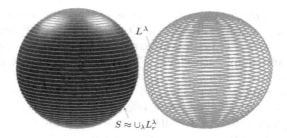

**Fig. 1.** Geodesic level curves analytically extraction on a sphere around a reference point

**Construction of the Radial Lines Curves.** We remind as well as the process of radial lines curves construction from a reference point $r$ of the surface $S$.

Like mentioned in [6], the radial curves represent a solution of the following system:

$$\begin{cases} \frac{dP(t)}{dt} = -\nabla U_r(P) \\ P(0) = r \\ \frac{dP(t)}{dt}\big|_{t=0} = \alpha \end{cases} \qquad (2)$$

where $P(t)$ is the geodesic path emanating from $r$ and following the opposite gradient $\nabla$ direction on $U_r$. Radial lines curves, denoted by $C^\alpha$, are therefore generated according to the angular direction $\alpha$ which can be arbitrary taken. Similar to geodesic level curves, a reconstruction of the surface $S$ can be approximated by $\cup_\alpha C^\alpha$, which is illustrated in Fig. 2.

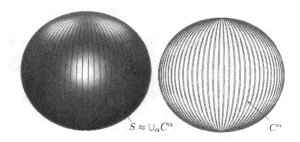

**Fig. 2.** Radial lines analytically extraction on a sphere from a reference point

**The Darcyan Representation.** Here we define Darcyan representation $D$ as the superposition of both $n$ geodesic level curves and $m$ radials lines curves relatively to a given point $r$.

$$D_{k,l}(r) = \left\{ L_r^{\lambda k} \bigcup C_r^{\alpha l}, 1 \leq k \leq n, 1 \leq l \leq m \right\}$$

The Fig. 3 illustrates the steps of Darcyan coordinate system construction. We propose to compute the curvature values on the intersection points between the radial line curves and the geodesic level ones of the Darcyan coordinate system. These intersection points are invariant under the $M(3)$ group. They are also ordered since each point on the surface is indexed by the geodesic level curve and the radial one to which it belongs.

## 2.2 Brief Recall of the Curvature Computation

First, we recall some useful facts for the curvature calculation on surface.

Let $X : (u, v) \in D \subset R^2 \rightarrow (x(u, v), y(u, v), z(u, v)) \in S \subset R^3$ be a parametrization of $S$.

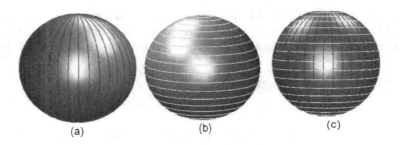

(a)                    (b)                    (c)

**Fig. 3.** Darcyan system reconstruction: Radial curves (a), geodesic level curves (b) and the superposition of both system of curves (c)

We consider $(u, v)$ the corresponding basis of the tangent plane to $S$ at a point $p = X(x_u, x_v)$. $N(p) = \frac{x_u \wedge x_v}{\|x_u \wedge x_v\|}$ is the normal vector to $S$ at $p$, according to a chosen orientation.

Therefore, the curvature expression is given using the following coefficients:

$$E = x_u.x_u, F = x_u.x_v, G = x_v.x_v, L = x_{uu}.\vec{N}, M = x_{uv}.\vec{N} \ and N = x_{vv} \vec{N}$$

$$K_G = \frac{LN - M^2}{EG - F^2}$$

$$K_M = \frac{EN - 2FM + GL}{2(EG - F^2)}$$

where $E$, $F$ and $G$ are the first fundamental coefficients, while $L$, $M$, $N$ are the second fundamental coefficients.

The quantities $K_G$ and $K_M$ are the Gaussian curvature and the Mean curvature $p = X(x_u, x_v)$ respectively.

Thus, the principal curvatures are derived from these expressions $K_G = k_{max}.K_{min}$ and $K_M = \frac{(k_{max} + k_{min})}{2}$.

$k_{max}$ and $k_{min}$ define the principal curvatures of the surface as, respectively, the maximal and the minimal curvature.

### 2.3    The Proposed Local Descriptor

We propose, here, a novel 3D shape descriptor which explores an intrinsic geometric property, principal curvatures fields on a local parametrization which is invariant under Euclidean motions.

The proposed descriptor relies on the computation of both principal maximal and minimal curvature field for a discrete point picked on a given geodesic level curve in the parametric representation, since these set of curves are invariant under motion group.

We propose to compute the mean of both principal maximal and minimal curvatures on the intersection points of each geodesic level curve. We denote $\overline{k^i_{max}}$ and $\overline{k^i_{min}}$ the mean of, respectively, $k_{max}$ and $k_{min}$ for the $i^{th}$

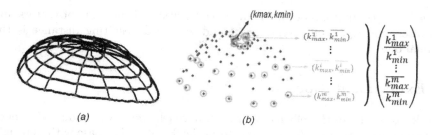

**Fig. 4.** Illustration of the proposed descriptor: (a) the Darcyan representation construction, (b) the vector of curvature fields computation and the obtained intersection points (in red) (Color figure online)

geodesic level curve. Thus the novel descriptor is defined by $\left\{ \overline{k_{max}^1}, \overline{k_{min}^1}, .., \overline{k_{max}^i}, \overline{k_{min}^i}, ..., \overline{k_{max}^m}, \overline{k_{min}^m} \right\}_{1 \le i \le m}$. The proposed descriptor is illustrated in Fig. 4.

## 3   3D Local Matching

Once we have extracted a set of interest points from shape extremities by appling an Average Geodesic Distance function (AGD) [12], we compute the proposed descriptor around each point of interest. Thus, we obtain vectors consisting on the mean of principal curvature field values.

After executing this process, the similarity between the acquired vectors is measured in term of $L_2$ distance. The full approach that we propose in shown

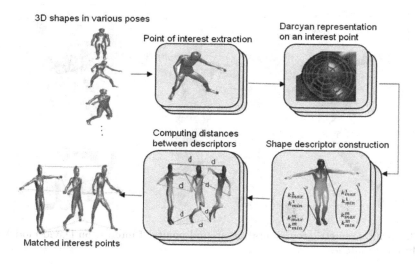

**Fig. 5.** 3D local correspondence approach

in Fig. 5. We compute, indeed, the distance between all the pairs of vectors and thereafter, correspondent points are found when the resulting distance is the minimal one.

## 4   Experimentation

In order to evaluate the efficiency of our approach, we have conducted experiments on 3D shapes in different poses from the TOSCA database with non rigid deformations.

### 4.1   The Used Database

We have conducted experiments on the high resolution TOSCA database, widely used in a variety of 3D shape correspondence approaches, which contains 80 meshes representing people and animals in a variety of poses. The meshes are grouped in 8 groups with common topology [1]. The reference points and the correspondence ground truth are provided in the evaluation benchmark proposed by [8].

### 4.2   Approximation on 3D Meshes

Since 3D surfaces are usually approximated by 3D meshes, we use the Fast Marching Method [9] for the geodesic distances computation. We, as well as, use Meyer et al. [11] algorithm for estimating the principal curvatures. Moreover, All the objects of the database need a scale normalization.

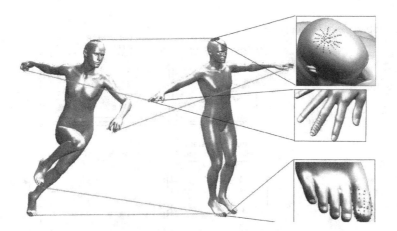

**Fig. 6.** Local parametric representation around points of interest on David model and the obtained matches

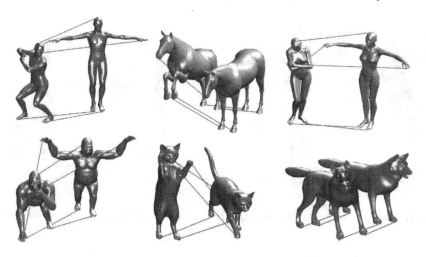

**Fig. 7.** Matching results for various models from the Tosca database

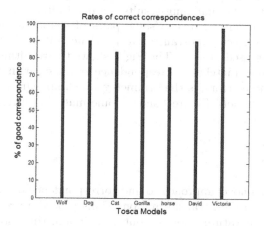

**Fig. 8.** Obtained correspondence rates for Tosca database

After being constructed, the sets of two curves require an interpolation step in order to increase their resolution and have more accuracy. Figure 6 illustrates local parametric representation around points of interest on the object David and the obtained matches.

**Fig. 9.** Incorrect matches (in blue) due to symmetric shapes (Color figure online)

### 4.3 Obtained Results

We evaluate our matching approach on different objects of the Tosca database. The Fig. 7 illustrates the matching results obtained after applying our process for finding correspondence between some pairs of shapes in various poses. In fact, our method involves important correspondence rates varying from 75% to 100% according the used objects. The Fig. 8 shows the resulting rates.

Beside the correct matches, correspondence errors exist in some cases. One of the factors of these errors, is the symmetry which appears in the majority of the used database models. Figure 9 shows some matching errors which are due to symmetric objects.

## 5 Conclusion

We have presented a novel approach to find correspondence between pairs of surfaces undergoing non-rigid transformations. This procedure relies on the mean values of principal curvature fields computation on a intrinsic local parametrization around reference points extracted on the extremities of the surface.

The performance of our local descriptor is evaluated on different models from Tosca Database. The first obtained results have shown the accuracy of our approach to establish correspondence among non-rigid shapes relying on the reached rates. Nevertheless, our matching process may lead to some errors and this is due to certain limitations such as symmetric shapes.

We intend, in the future works, to handle the problem of symmetry. In other hand, we aim to optimize the resolution of our intrinsic descriptor by finding the optimal number of the geodesic levels curves and the radial lines curves. We also intend to perform the experimentation on others 3D non-rigid databases with different properties.

# References

1. Bronstein, A.M., Bronstein, M.M.: Regularized partial matching of rigid shapes. In: Forsyth, D., Torr, P., Zisserman, A. (eds.) ECCV 2008. LNCS, vol. 5303, pp. 143–154. Springer, Heidelberg (2008). doi:10.1007/978-3-540-88688-4_11
2. Bronstein, A.M., Bronstein, M.M., Kimmel, R.: Generalized multidimensional scaling: a framework for isometry-invariant partial surface matching. Proc. Nat. Acad. Sci. U.S.A. **103**(5), 1168–1172 (2006)
3. Bronstein, A.M., Bronstein, M.M., Kimmel, R., Mahmoudi, M., Guillermo, S.: A Gromov-Hausdorff framework with diffusion geometry for topologically-robust non-rigid shape matching, pp. 612–626 (2009)
4. Cohen, L., Kimmel, R.: Global minimum for active contour models. Int. J. Comput. Vis. **24**(1), 57–78 (1997)
5. Funkhouser, T., Shilane, P.: Partial matching of 3D shapes with priority-driven search. In: Proceedings of the Fourth Eurographics Symposium on Geometry Processing, pp. 131–142 (2006)
6. Gadacha, W., Ghorbel, F.: A stable and accurate multi-reference representation for surfaces of R3: application to 3D faces description. In: IEEE International Conference on Automatic Face and Gesture Recognition (FG 2013), Shanghai, China (2013)
7. Jiang, L., Zhang, X., Zhang, G.: Partial shape matching of 3D models based on the Laplace-Beltrami operator eigen function. J. Multimed. **8**(6), 655–661 (2013)
8. Kim, V.G., Lipman, Y., Funkhouser, T.: Blended intrinsic maps. ACM Trans. Graph. **30**(4), 1 (2011)
9. Kimmel, R., Sethian, J.A.: Computing geodesic paths on manifolds. Proc. Nat. Acad. Sci. U.S.A. **95**(15), 8431–8435 (1998)
10. Lähner, Z., Rodolà, E., Schmidt, F.R., Bronstein, M.M., Cremers, D.: Efficient Globally optimal 2D-to-3D deformable shape matching (2016)
11. Meyer, M., Desbrun, M., Schroder, P., Barr, A.: Discrete differential-geometry operators for triangulated 2-manifolds. In: Hege, H.C., Polthier, K. (eds.) Visualization and Mathematics III. Mathematics and Visualization (2002)
12. Hilaga, M., Shinagawa, Y., Kohmura, T., Kunii, T.: Topology matching for fully automatic similarity estimation of 3D shapes. Proc. SIGGRAPH **32**(6), 203–212 (2001)
13. Ovsjanikov, M., Mérigot, Q., Mémoli, F., Guibas, L.: One point isometric matching with the heat kernel. In: Eurographics Symposium on Geometry Processing, vol. 29, no. 5, pp. 1555–1564 (2010)
14. Sahillioğlu, Y., Yemez, Y.: Minimum-distortion isometric shape correspondence using EM algorithm. IEEE Trans. Pattern Anal. Mach. Intell. **34**(11), 2203–2215 (2012)
15. Sahillioğlu, Y., Yemez, Y.: Partial 3-D correspondence from shape extremities. Comput. Graph. Forum **33**(6), 63–76 (2014)
16. Sahillioglu,Y., Yemez, Y.: 3D shape correspondence by isometry-driven greedy optimization. In: Proceedings of the IEEE Computer Society Conference on Computer Vision and Pattern Recognition, pp. 453–458 (2010)
17. Thomas, F., Yaron, L.: Mobius Voting For Surface Correspondence. World (2009)
18. Thompson, D.: On Growth and Form. University Press, Cambridge (1917)
19. Van Kaick, O., Zhang, H., Hamarneh, G.: Bilateral maps for partial matching. Comput. Graph. Forum **32**(6), 189–200 (2013)

20. Van Kaick, O., Zhang, H., Hamarneh, G., Cohen-Or, D.: A survey on shape correspondence. Comput. Graph. Forum **xx**, 1–23 (2010)
21. Zhang, H., Sheffer, A., Cohen-Or, D., Zhou, Q., Van Kaick, O., Tagliasacchi, A.: Deformation-driven shape correspondence. In: Eurographics Symposium on Geometry Processing, vol. 27, no. 5, pp. 1431–1439 (2008)

# A Novel Robust Statistical
# Watermarking of 3D Meshes

Nassima Medimegh[1]($\boxtimes$), Samir Belaid[1]($\boxtimes$), Mohamed Atri[1]($\boxtimes$),
and Naoufel Werghi[2]($\boxtimes$)

[1] University of Monastir, Monastir, Tunisia
Medimegh_Nassima@yahoo.fr, {Samir.Belaid,Mohamed.Atri}@fsm.rnu.tn
[2] Khalifa University, Sharjah, UAE
Naoufel.Werghi@kustar.ac.ae

**Abstract.** In this paper, we present a novel robust blind 3D mesh watermarking approach. We embed signature bits into the vertex norms distribution. At first, the robust source locations are extracted by using a salient point detector, based on the Auto Diffusion Function (ADF). Afterwards, the mesh is segmented into different regions according to the detected salient points. Then, the same watermark bits are embedded statistically into each region. The experimental results show the robustness of our method against cropping and other common attacks. Due to the stability of salient points, we can retrieve the watermarked region and extract the watermark. In addition, the performance of our method is also demonstrated on the minimal surface distortion in the embedding process.

**Keywords:** 3D watermarking · Salient points · Fast marching method · Statistical method

## 1 Introduction

Digital watermarking is a technique used to add information to various media such as text, audio, image or video in order to protect either the copyright or the integrity of the digital content. In fact, copyright protection applications require a robust watermarking. This type of watermarking is employed to prove the proprietor of the digital content and to prohibit its illegal use. While fragile watermarking is used to protect the integrity of media from an unauthorized processing and to detect an eventual local or global manipulation. Most types of attacks are geometric attacks (geometric transformation, scaling, smoothing, noise addition) and topological attacks (mesh simplification, subdivision, cropping). According to the original model used or not in the extraction process, watermarking can be blind or non-blind.

Nowadays, three-dimensional meshes have been used more and more in many sectors. This brought awareness on the importance of their intellectual property protection and authentication problems to study the digital watermarking technique for 3D triangular mesh models.

© Springer International Publishing AG 2017
B. Ben Amor et al. (Eds.): RFMI 2016, CCIS 684, pp. 27–38, 2017.
DOI: 10.1007/978-3-319-60654-5_3

Since ohbuchi et al. [1] proposed the first 3D mesh watermarking, many watermark algorithms are developed and tried to ameliorate the performance. Generally speaking, the 3D watermarking can be applied in spatial or spectral domain. Indeed, in spatial domain, the geometry or the topology of the mesh is generally changed. Whereas, the spectral watermarking is based on the modification of the transformed components (spectral and multiresolution).

This paper focuses on the robust and blind 3D mesh watermarking. Our main objective is to achieve an imperceptible and robust watermarking against different attacks. The proposed approach is based on two main steps: the mesh segmentation, and the statistical watermark embedding in each segmented region.

The rest of this article is constituted as follows: in Sect. 2, we will present some existing methods. Some fundamental concepts used in our method are introduced in Sect. 3. Section 4 describes the watermark embedding and extraction method in detail. Before concluding, experiments and obtained results are given in Sect. 5.

## 2    Related Works

As mentioned before, 3D watermark algorithm performed in the spatial and transform domains. A general survey of 3D watermarking is presented in [2].

Watermarking methods can also be a statistical or deterministic methods. As experiments prove that statistical method are more robust then the other classes in this paper, we focus on this type which extract signature by a statistical test.

Zafeiriou et al. [3] proposed the Principal Object Axis (POA) method whereby the signature is embedded by modifying the $\rho$ component of the vertex. Then with a second method Sectional Principal Object Axis (SPOA), they have displaced the set of vertices having the coordinate $\theta$ within a specific ranges.

Two blind and robust statistical methods are proposed by Cho et al. [4]. They changed the mean or the variance of the histogram distribution of the vertices norms depending on the watermark bit. Despite the algorithm is robust against the distortion attacks such as additive noise, smoothing and mesh simplification, it causes visible artifacts on the 3D models surface. Hu et al. [5] proposed a similar statistical method by using quadratic programming in order to minimize the mean square error between the original mesh and the watermarked mesh. This method is more robust to Gaussian noise, compared with Cho's method. However, it has difficulties with large meshes. Alface et al. [6] embedded a watermark in a local neighborhood by applying locally Cho's method [4].

Luo and Bors [7] modified the statistics of the distances then applied the Quadratic Selective vertex placement scheme in order to find the suitable position of some vertex. In other statistical method [8], they embedded signature into geodesic distance distribution by using the Fast Marching method (FMM). Bors and Luo [9] proposed a new statistical approach by minimizing the 3D object surface distortion. They used the Levenberg-Marquardt optimization method for spherical coordinates of vertices.

Nakazawa et al. [10] embedded statistically the watermark to some regions after segmentation of mesh.

# 3  Preliminaries

In this section, we introduce some fundamental concepts used in our method. We start with a brief description of the mesh structure. Then we introduce the concept of Laplace-Beltrami operator. We also present the Heat Kernel Diffusion process.

## 3.1  Structure of Representation

In general, a mesh is a set of polygons. It is defined by three elements: vertices, faces and edges. It contains two informations: a geometric information which defines essentially by the vertex coordinates and a connectivity information (topological) that reflects the adjacency relationship between vertex. The neighborhood of a vertex is the set of vertices that are directly connected to this vertex by an edge.

There are other mesh representations such as cloud of points and parametric surface. However, the triangular mesh is the most used in watermarking than the other structures.

## 3.2  Laplace-Beltrami Operator

In the case of a manifold triangular mesh M, the Laplace-Beltrami operator $\Delta$ has a non-trivial solution. The definition of the discrete Laplace-Beltrami by the Finite Element Method [11] is:

$$-Qh = \lambda Dh \tag{1}$$

Where h is the eigenfunction associated to the eigenvalue $\lambda$.
D is called the lumped matrix defined by:

$$D_{ii} = \frac{1}{3} \sum_{t \in N_t(i)} |t| \tag{2}$$

Where $N_t(i)$ is the ensemble of set-facets from vertex $v_i$
And Q is the stiffness matrix is given as follows:

$$Q_{ij} = \begin{cases} Q_{i,j} = \frac{\cot(\beta_{i,j}) + \cot(\beta'_{i,j})}{2} \\ Q_{i,i} = -\sum_j Q_{i,j} \end{cases} \tag{3}$$

Where the two angles $\beta_{i,j}$ and $\beta'_{i,j}$ appearing in this formula are the opposite angles to the edge $v_i v_j$ .

### 3.3   The Heat Kernel

The heat equation describes the distribution of heat in a given region over time. In an unstable condition, the heat diffusion process represents the progress of a function onto a surface. It is patterned by the method Heat Kernel HK. $K_t(x,y)$ is presented as the probability that the point y is achieved from the point x at time t. One way to solve the equation of the HK is to use the basis of eigenfunctions correspond to different eigenvalues of Laplace-Beltrami operator. The heat kernel has the following spectral decomposition:

$$K_t(x,y) = \sum_{n=0}^{\infty} \exp\left(-\lambda_n t\right) h_n(x) h_n(y) \tag{4}$$

The heat kernel $K_t(x,y)$ is symmetric and invariant under isometric deformations due to the invariance of the Laplace-Beltrami operator.

### 3.4   Geodesic Distance

The geodesic distance is determined by the length of the shortest path joining two locations on the object mesh surface M. In the following, we present the concepts of calculating the geodesic distance on manifolds.

Let us consider a curve which joins two different points x and y located on the object surface. In fact, the geodesic distance $T_o$(x,y) between the endpoints x and y is the shortest length over all continuous paths determined by the geodesic curve $\gamma$(t)(t).

$$T_o(x,y) = \min_{\gamma(t)\in o} \int_0^P \sqrt{\gamma'(t)^T H(\gamma(t))\gamma'(t)} dt \tag{5}$$

where $\gamma(0)$=x, $\gamma(P)$=y, $\gamma$'(t) represents the local derivative of the parametric curve, and H(.) is an intrinsic metric, considered as H(.)=1.

As the geodesic distance is an integrative function, it resists to smoothing, noise adding and resampling.

## 4   The Proposed Method

The proposed method embeds watermark statistically and locally into 3D triangular mesh model by modifying the component spherical $\rho$ of vertex according to assigned watermark bit and distribution of $\rho$. Figure 1 depicts the different stages of the embedding process.

Robust feature points are achieved using a salient point detector. Then, the 3D model is divided into region. In the last stage, the watermark is embedded into each regions. Here, we employed the embedding technique of Cho et al. [4]. The extraction is performed without any need to the original model.

The algorithm for watermark embedding and extraction is described in detail in the following subsection.

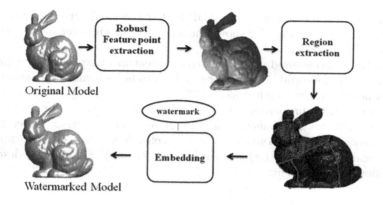

**Fig. 1.** Watermark embedding process

## 4.1 Feature Point Extraction

In order to extract a robust source location from the mesh model, we used the salient point detector proposed by Haj-Mohamed and Belaid [12].

Salient points must have a distinct locality, and must be stable at all cases of a model. In addition, those points are not altered by scaling, rotation, additive noise, articulation, and deformation. In this context, Haj-Mohamed and Belaid [12] presented an unsupervised and automatic 3D salient point detector founded on Auto Diffusion Function ADF introduced by Gbal et al. [13].

This scalar function is defined as a linear sequence of the Laplace-Beltrami Operator eigenfunctions. To supervise the number and the capacity of the extracted features, just we varied the parameter t.

$$ADF_{\frac{t}{\lambda_2}} = K(x, x)$$
$$= \sum_i \exp\left(-t\frac{\lambda_t}{\lambda_2}\right) h_i^2(x) \tag{6}$$

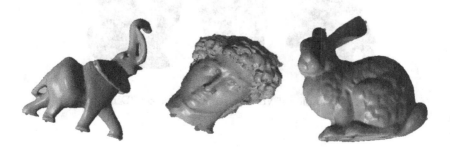

**Fig. 2.** Example of detected salient points.

The important advantage of the ADF is that the local extrema prove to be natural interest points. Figure 2 shows the feature points in some mesh extracted using the ADF function.

It has been also showed that the extracted features have important properties: isometry or non-rigid deformations and micro holes invariant, insensitive to Gaussian noise and Laplacian smooth [12].

In this paper, we propose a novel method consists on embedding watermark locally. The key idea is to benefit from the references locations extracted to segment the object surface into geodesic Voronoi cells. So for that, we segmented the mesh in local region using the fast marching algorithm FMM which is described in the following subsection.

## 4.2   Geodesic Voronoi Strip Genaration

We use the fast marching algorithm to compute the geodesic distance from a set of input feature points extracted in the previous step, and then to segment the surface into geodesic Voronoi cells. In this case the Geodesic Voronoi Diagram is discriped in the following.

Let $P = p_1, p_2,..., p_m$ be an ensemble of points on mesh M. The Voronoi cell $VC(p_i)$ of the site pi includes all points whose distance to $p_i$ is less than or equal to their distance to any other site [14], i.e., $VC(p_i) = \{q \in M | d(p_i, q) \leq d(p_j, q),$ for all i $\neq$ j$\}$. So, the geodesic Voronoi diagram (GVD) of P is the union of all Voronoi cells, $GVD(P) = \{VC(p_1), VC(p_2), ... , VC(p_m)\}$.

The Voronoi segmentation is shown using colors on some 3D models in Fig. 3, where the salient points is indicated by a small red and the regions extracted are colored.

**Fig. 3.** Examples of Voronoi segmentation on 3D models.

## 4.3  Watermark Embedding Process

In the following, we consider the same notations and terminologies as in [4] for the raison that our research work follows theirs. Here, we employed the embedding technique of Cho et al. [4]. In this method, watermark is embedded at the cadence of one bit per bin in each region. Here, the mean value of each set is modified depending on the assigned signature bit. In fact, by using a histogram mapping function, we modify norms of the vertex in each bin. Afterwards, we will detail the embedding process where is summarized in Fig. 4 which depicts the different stages.

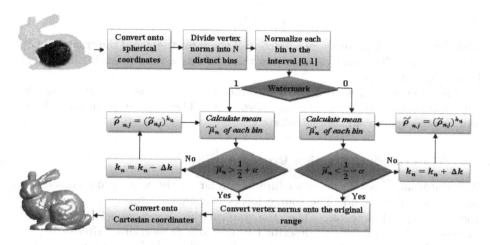

**Fig. 4.** Watermark embedding process for Cho's method.

Let us assume that we want to insert N bits of a watermark $w$ into the vertices $v_i$ of 3D mesh. Firstly, the Cartesian coordinates of a given point $v_i$ $(x_i, y_i, z_i)$ in the mesh are converted into spherical coordinates $(\rho_i, \theta_i, \varphi_i)$. Secondly, the vertex norms are distributed over N distinct bins based on their magnitude. As we used one bin to insert just one bit watermark, we get a total of N embedding bits per selected region in the mesh. Then, we get the vertex norms $\rho_{max}$ and $\rho_{min}$ by:

$$\rho_{n,min} = \rho_{min} + \frac{\rho_{max} - \rho_{min}}{N}.n \tag{7}$$

$$\rho_{n,max} = \rho_{min} + \frac{\rho_{max} - \rho_{min}}{N}.(n+1) \tag{8}$$

in which the bin $B_n$ expressed as:

$$B_n = \{\rho_{n,j} | \rho_{n,min} < \rho_{n,j} < \rho_{n,max}\} \tag{9}$$

where the $\rho_{n,min}$ and $\rho_{n,max}$ are lower and upper boundaries of the nth bin, and $\rho_{n,j}$ is the jth vertex norm in the nth bin.

The third step is to give the normalization of the vertex norms of the bin $B_n$ to the interval $[0, 1]$ as:

$$\tilde{\rho}_{n,j} = \frac{\rho_{n,j} - \rho_{n,min}}{\rho_{n,max} - \rho_{n,min}} \tag{10}$$

where $\tilde{\rho}_{n,j}$ is the normalized jth vertex norm in the nth bin.

The watermarking realised here by perturbing $\tilde{\rho}_{n,j}$, to make the mean value of each bin via transforming vertex norms shifted via the histogram mapping function. So that the mean of the histogram is moved to a specific area according to the watermarking bit to be embedded as follows:

$$\tilde{\mu}_n = \begin{cases} \frac{1}{2} + \alpha & if\ w_n = +1 \\ \frac{1}{2} - \alpha & if\ w_n = 0 \end{cases} \tag{11}$$

whith the strength parameters $\alpha$ to control the robustness and the transparency of watermark. The iterative algorithm that embeds a watermark bit into a bin is described in Fig. 4.

The final step of watermark embedding process consist on converting the spherical coordinates to Cartesian coordinates. These six steps of embedding will be repeated for all regions selected.

## 4.4    Watermark Extraction Process

The watermark extraction procedure is similar to the watermark embedding process and it is blind. The source location should be firstly detected using the salient point detector. Then, the watermarked mesh model is segmented in geodesic Voronoi cells as described in previous subsections. Afterwards, we employed the watermarking technique of Cho et al. [4]. In fact, the watermarked mesh object is converted to spherical coordinates. Here, the vertex norms in each region are classified into bins and mapped onto the normalized range between 0 and 1, and then, the mean of each bin is calculated and compared with $1/2$. The bit of the nth bin is obtained according to:

$$w'_n = \begin{cases} +1 & if\ \tilde{\mu}" > \frac{1}{2} \\ 0 & if\ \tilde{\mu}" < \frac{1}{2} \end{cases} \tag{12}$$

## 5    Experimental Results

The proposed statistical watermarking method was applied on mesh objects used usually in 3D watermarking application. In the following we provide the results of our method on the mesh objects: Bunny, Cow, Elephant, Fandisk and David. These original models used are displayed in Fig. 5, and their mesh characteristics are presented in Table 1.

**Fig. 5.** 3D objects used in the experiments.

**Table 1.** Characteristics of the mesh models

| Models | Facets | Vertices |
|---|---|---|
| Bunny | 25736 | 12942 |
| Cow | 9018 | 9511 |
| Elephant | 24950 | 12471 |
| David | 25640 | 12865 |
| Fandisk | 25892 | 12948 |

## 5.1 Mesh Distortion

The good visual imperceptibility of the watermark is one of property of a watermarking method. For mesh models, Haussdorff distance HD has been adopted to measure the quality distortion resulted by the watermarking. Based on Metro [15], we measure the HD between the original mesh model and marked one with the following formula:

$$HD = max\{h(M_1), h(M_2)\} \tag{13}$$

where $M_1 = (V, V')$ and $M_2 = (V', V)$, ($V$ and $V'$ represent respectively the original mesh and watermarked mesh).

$h(M_1) = max\{min(d(a,V'))\}$, $a$ in $V$, $h(M_2) = max\{min(d(b,V))\}$, $b$ in $V'$.

We evaluated the visual effect of the mesh alterations produced by watermarking. Figure 6 displays the visual effects of the proposed watermarking methods. No differences can be observed between the original and the marked object. Furthermore, the low result of the calculated HD proves the good invisibility. As can be observed from Table 2, We notice that the object surface distortion produced by the proposed watermarking method is lower than that introduced by Cho's.

## 5.2 Robustness

In the following, we evaluate the watermark robustness to various attacks. The robustness is measured by calculating the correlation between the original and the extracted signature.

**Fig. 6.** Marked 3D objects.

**Table 2.** Watermarked object distortion

|  | Bunny | | Cow | | Elephant | | David | | Fandisk | |
|---|---|---|---|---|---|---|---|---|---|---|
|  | Our | Cho | Our | Cho | Our | Cho | Our | Cho | Our | Cho |
| HD ($\times 10^{-3}$) | 0.73 | 0.74 | 5.8 | 5.8 | 3.5 | 7.8 | 5.2 | 5.2 | 4.7 | 2.6 |

$$corr = \frac{\sum_{n=0}^{N-1}(w_n - \bar{w})(w_n' - \bar{w}')}{\sqrt{\sum_{n=0}^{N-1}(w_n - \bar{w})^2 \times \sum_{n=0}^{N-1}(w_n' - \bar{w}')^2}} \tag{14}$$

where $\bar{w}$ is the average of the signature and the correlation corr in $[-1, 1]$.

To evaluate the robustness, different attacks are tested. Among these attacks, such as RST attacks (translation, scaling and rotation), Laplacian smoothing, additive noise, mesh quadratic metric simplification, and cropping, we can extract watermark. Figure 7 gives a comparison result between our approach and Cho's watermarking algorithm [4]. Our method generally provides similar results of robustness as Cho et al. [4].

We note that, for every tested model, we extract the signature after attacks. Precisely in the case of cropping, although there were some region are removed, the watermark could be extracted from the other patches.

In addition, we compute the number of regions where we detect the watermark after different attacks. So, for every tested model, we extract the signature from many regions.

As can be observed from the curves in Fig. 7, our method provide better results for the smoothing attack and slightly better for the other attacks. Furthermore, our approach resist to cropping attack. Embedding the watermark in many regions across the 3D mesh model facilitate the detection after a cropping.

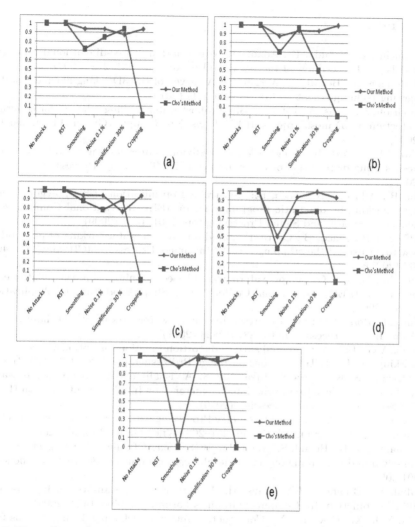

**Fig. 7.** Plots showing the robustness against attacks. (a) Bunny. (b) Cow. (c) Elephant. (d) David. (e) Fandisk.

## 6    Conclusion

In this paper, we presented a novel statistical 3D watermarking method. In fact, we define the salient points used in segmentation of 3D models into geodesic Voronoi region. A statistical method is used to embed signature by changing the mean of distributions of vertex norms. We have improved the robustness of our blind watermarking scheme to cropping attacks.

# References

1. Ohbuchi, R., Masuda, H., Aono, M.: Watermarking three-dimensional polygonal models. In: Proceedings of the ACM Multimedia, pp. 261–272 (1997)
2. Medimegh, N., Belaid, S., Werghi, N.: A survey of the 3D triangular mesh watermarking techniques. Int. J. Multimedia **1**(1) (2015)
3. Zafeiriou, S., Tefas, A., Pitas, I.: Blind robust watermarking schemes for copyright protection of 3D mesh objects. IEEE Trans. Vis. Comput. Graph. **11**(5), 596–607 (2005)
4. Cho, J.W., Prost, R., Jung, H.Y.: An oblivious watermarking for 3-D polygonal meshes using distribution of vertex norms. IEEE Trans. Sig. Process. **55**, 142–155 (2007)
5. Hu, R., Alface, P.R., Macq, B.: Constrained optimisation of 3D polygonal mesh watermarking by quadratic programming. In: IEEE International Conference on Acoustics, Speech and Signal Processing, pp. 1501–1504 (2009)
6. Alface, P.R., Macq, B., Cayre, F.: Blind and robust watermarking of 3D models: how to withstand the cropping attack? In: Image Processing, ICIP 2007, pp. 465–468 (2007)
7. Luo, M., Bors, A.G.: Shape watermarking based on minimizing the quadric error metric. In: IEEE International Conference on Shape Modeling and Applications (SMI) (2009)
8. Luo, M., Bors, A.G.: Surface-preserving robust watermarking of 3-D shapes. IEEE Trans. Image Process. **20**(10), 2813–2826 (2011)
9. Bors, A.G., Luo, M.: Minimal surface distortion function for optimizing 3D watermarking. In: IEEE Transactions on Image Processing, pp. 1822–1835 (2013)
10. Nakazawa, S., Kasahara, S., Takahashi, S.: A visually-enhanced approach to watermarking 3D models. In: International Conference on Intelligent Information Hiding and Multimedia Signal Processing (2010)
11. Vallet, B., Levy, B.: Spectral Geometry Processing with Manifold Harmonics, Computer Graphics Forum, vol. 27, pp. 251–260. Wiley Online Library (2008)
12. Mohamed, H.H., Belaid, S.: Algorithm BOSS (Bag-Of-Salient local Spectrums) for non-rigid and partial 3D object retrieval. Neurocomputing. doi:10.1016/j.neucom. 2015.05.045
13. Gbal, K., Brentzen, J.A., Aans, H., Larsen, R.: Shape analysis using the auto diffusion function. In: Computer Graphics Forum, pp. 1405–1413 (2009)
14. Liu, Y.J., Xu, C.X., He, Y.: Constructing Intrinsic Delaunay Triangulations from the Dual of Geodesic Voronoi Diagrams arXiv:1505.05590[cs.CG]
15. Cignoni, P., Rocchini, C., Scopigno, R.: Metro: measuring error on simplied surfaces. In: Computer Graphic Forum, vol. 17, pp. 167–174 (1998)

# Face Analysis and Recognition

# Shape Analysis Based Anti-spoofing 3D Face Recognition with Mask Attacks

Yinhang Tang[✉] and Liming Chen

Université de Lyon, Ecole Centrale de Lyon,
LIRIS laboratory UMR CNRS 5205, 69134 Lyon, France
tang-yinhang@doctorant.ec-lyon.fr

**Abstract.** With the growth of face recognition, the spoofing mask attacks attract more attention in biometrics research area. In recent years, the countermeasures based on the texture and depth image against spoofing mask attacks have been reported, but the research based on 3D meshed sample has not been studied yet. In this paper, we propose to apply 3D shape analysis based on principal curvature measures to describe the meshed facial surface. Meanwhile, a verification protocol based on this feature descriptor is designed to verify person identity and to evaluate the anti-spoofing performance on Morpho database. Furthermore, for simulating a real-life testing scenario, FRGCv2 database is enrolled as an extension of face scans to augment the ratio of genuine face samples to fraud mask samples. The experimental results show that our system can guarantee a high verification rate for genuine faces and the satisfactory anti-spoofing performance against spoofing mask attacks in parallel.

## 1 Introduction

As the most significant biometric trait of human beings, the human face has been widely used for human identification and verification in the scientific research and the real-world application. The facial data acquisition method is natural, non-intrusive and contactless, which is friendly to accept in social activities [11,12]. With the development of 3D scanner, 3D printer, Virtual Reality (VR) and Augmented Reality (AR), capturing and reconstructing 3D samples become more convenient in daily-life [30]. Meanwhile, 2D and 3D face recognition have been applied widely in the criminal investigation, the access control, the frontier inspection and the bank service. Even though the techniques of the face recognition have been widely studied in biometrics research area [1,2,9,19,26,36] and many state-of-the-arts have been reported in many publications, the spoofing attacks against face recognition systems is a potential threat to biometric application.

Spoofing attack is defined as an intrusive act of deceiving a biometric system by presenting a fake evidence or a copied biometric trait to obtain a valid authentication [28]. By using photographs or videos captured in distance or collected via internet, the attacker can easily achieve the facial information of a

© Springer International Publishing AG 2017
B. Ben Amor et al. (Eds.): RFMI 2016, CCIS 684, pp. 41–55, 2017.
DOI: 10.1007/978-3-319-60654-5_4

valid user registered in a face recognition system. Then the attacker shows the fake photograph printed on paper or the recorded video displayed on a tablet for attempting to get access in the system. Furthermore, since few yeas ago, a social public website "Thats My Face"[1] started to provide the wearable 3D mask manufacturing service with only one frontal photo (another side-view photo is asked as option). It further reduced the difficulty of attacker's deception by wearing such a 3D printed mask. The simplicity and the convenience of the acquirement and the manufacture of the 2D/3D face data, which should be the advantages of the face recognition, become gradually the jeopardy and the calamity to the reliability and the stability of the face recognition system.

Due to the vulnerability of face recognition systems, many papers have been published on countermeasure studies, and the reported experimental results showed that the corresponding methods are sufficient and efficient. Among the published works, liveness detection [3,15,29,39], motion detection [5,14] and texture analysis [20,25] are three principal categories of anti-spoofing methods [6,7] against photo- and video-based spoofing attacks. However, with the help of the improvement of 3D manufacture technology, the easily obtained high-quality 3D masks introduce new challenge to anti-spoofing research. Morpho database[2] and 3DMAD database [6], including 2D, 2.5D and 3D face samples of genuine person and imposters wearing 3D mask, were constructed for simulating this mask intrusion. Kose et al. proposed the countermeasure based on the fusion of the information extracted from texture and depth images, and tested it on the Morpho database [16–18]. Erdogmus et al. evaluated various LBP based countermeasures on texture images in [6,7]. All of their works reported that the texture information can be essential discriminative characters to distinguish real faces and masks. However, the shape analysis based approach, as an important kind of methods in 3D face recognition, has not been discussed and studied in their works. To fill this gap, in this paper, we aims to evaluate the anti-spoofing performance of this kind of method.

The general technique of the shape analysis based 3D face recognition is to utilize the geometric attributes to describe and characterize the facial surface precisely. Geometry attributes, including principal curvatures, Gaussian curvature, mean curvature and their variations (e.g. shape index), have been commonly used to 3D face representation [21,22,35], keypoints location [24,26,38] and 3D facial feature descriptor generation [8,19,31]. In this study, we exploit "curvature measures" developed by [27,32,33] based on concept of normal cycle [4] to extract shape information of discrete surface (e.g. 3D meshed face), and design the corresponding facial description and recognition framework. This triangle mesh based geometric feature can highlight the micro-shape dissimilarity between genuine faces and manufactured masks, which leaves us more opportunities to distinguish them.

Besides, in the real-life scenario, comparing to the verification cases with genuine face samples, the spoofing mask attacks appear more rarely, which are

---

[1] http://www.thatsmyface.com.
[2] http://www.morpho.com.

regarded as an exceptional testing case. In order to simulate such a scenario, we firstly propose to combine Morpho database and FRGCv2 database [1], the largest public 3D face database, to appraise the anti-spoofing performance. The gallery set and the genuine probe set are formed by the genuine face samples coming from both Morpho database and FRGCv2 database, while the testing fake face scans from Morpho database are 5% and 1% of the scale of the whole probe set. This performance evaluation scenario corresponds better to the real-life case. Meanwhile, the conventional discriminative power evaluation and anti-spoofing performance test on Morpho database are also reported in the experiment part.

The rest of this paper is organized as follows. Section 2 gives a brief review of related anti-spoofing face recognition works. Section 3 presents principal curvature measures estimation method and the related shape analysis based facial feature descriptor. Section 4 shows the experimental results in several scenarios, and Sect. 5 finally concludes the paper.

## 2 Related Work

In the history of mask anti-spoofing research, the work of Kim *et al.* [13] can be regarded as the first published one. Due to the difference of the reflectance between face skin and materials used to manufacture mask, their work aimed to analyse the distribution of albedo values for illumination at various wavelengths. Based on Fisher's linear discriminant, they selected a 2D feature vector consisting of radiance measurement to be the classification criteria in visual and NIR spectrum (685 and 850 nm respectively). Similarly, Zhang *et al.* published their mask detection countermeasure based on multi-spectral analysis in [39]. They claimed to abandon visual face image, but to analyse multi-spectral images captured in two discriminative wavelengths of illumination (850 and 1450 nm). They measured the albedo curves of different materials and trained SVM classifier to distinguish real face and mask. Even though these two papers above are effective in mask distinguishing, they haven't resolve the anti-spoofing problem. Besides an extra multi-spectral capturing device is obligatory in their defence system.

Lately, Kose *et al.* reported their anti-spoofing works based on texture and depth information in 2D and 2.5D images from Morpho Database. Three baseline face recognition algorithms are tested and an anti-spoofing mask attack related experimental strategy is mentioned in [17]. Furthermore, they extracted LBP features in color and depth image and trained the linear SVM classifier to determine whether the input sample is genuine or fake in [16]. Then in order to combine the advantage of color and depth image in anti-spoofing mask attack, two feature and score level fusions were proposed in [18]. They concluded that texture analysis is a effective method for developing a countermeasure. Similarly, Erdogmus *et al.* introduced their 3D Mask Attack Database (3DMAD) and anti-spoofing countermeasures based on three extended LBP algorithms in [6]. Some more comparative anti-spoofing experimental results on Morpho and 3DMAD databases were reported in [7]. Even though the works presented above introduced great countermeasures against 3D mask attacks, all of their main methods are designed based on 2D images projected from 3D face scans. Moreover their evaluation of the anti-spoofing capability of 3D face samples is constructed relying on Thin Plate Spline (TPS) warping parameters and Iterative Closest Point (ICP) methods. But 3D shape based geometric attributes, as potent characters in 3D shape analysis, haven't been evaluated their anti-spoofing potentiality. Besides, all their experiments are limited into the mask attack related database which doesn't includes not enough genuine face scans to test. In this paper, we have two main purposes to fill the gaps:

- We take full advantage of *principal curvature measures* based on asymptotic cone theory to design a facial descriptor, and evaluate the potential anti-spoofing performance of the facial descriptor for 3D meshed face scans.
- For simulating a real-life recognition case, we firstly attempt to combine a scale limited genuine-imposter combined database (*i.e.* Morpho database) to a large genuine face scans database (*i.e.* FRGCv2 database), and perform comprehensive experimental scenarios.

## 3   Principal Curvature Measures Based 3D Face Recognition Scheme

The shape analysis is regarded as an important kind of 3D face recognition methods. Among this branch of approaches, both a precise estimated geometric attribute and a proper related facial feature descriptor are significant to represent and describe the shape of facial surface. In this section, we will introduce our principal curvature measures estimation method and their related 3D face recognition scheme, which meet these two demands of this kind of method. Meanshile, because the surface of manufactured mask is smoother than real face which is stated in [17,18], our proposed descriptor, which can highlight the dissimilarity of the minor shape between genuine faces and masks, is capable to verify the liveness of the testing samples. The pipeline of our proposed method includes 4 steps: principal curvature measures estimation, 3D keypoint detection, 3D keypoint feature description and 3D keypoint matching.

### 3.1   Principal Curvature Measures Estimation

In general speaking, principal curvatures are the most basic but fundamental geometrical attributes in differential geometry. They are defined regularly as below: Suppose a point $p$ locating on a smooth oriented surface $S$, its principal curvatures $\lambda_{1_p}$ and $\lambda_{2_p}$ are estimated as the eigenvalue sets of the corresponding second fundamental form $h$ ($q$ is the quadratic form associated to $h$). $\lambda_{1_p}$ and $\lambda_{2_p}$ can describe the local bending information around $p$ of $S$. Remark that this definition is coherent because of the smoothness of surface $S$. However, 3D face sample is commonly recorded as the triangle mesh, which is continuous but piecewise smooth. It makes the conventional curvature estimation method unsuitable here. A possible solution proposed and demonstrated in [32,33] is to generalize the definition of curvatures to the discrete surface and to replace functions by measures. Here we will present the generalization from the smooth surface case to the discrete surface case. Please refer [32,33] for a comprehensive and detailed introduction of the generalization.

**Principal Curvature Measures of Smooth Surfaces.** Inspiered to [32,33], the second fundamental form $h$ and its associated quadratic form $q$ can be generalized to a measure on a smooth surface $S$ on $\mathbb{E}^3$. Suppose that any Borel subset $B$ of $\mathbb{E}^3$ and any vector field $X$ of $\mathbb{E}^3$, the definition of $\overline{h}_B$ and $\overline{q}_B$ are:

$$\overline{h}_B(X,X) = \int_{S \cap B} h_p(pr_{T_pS}X_p, pr_{T_pS}X_p)dp,$$

$$\overline{q}_B(X) = \int_{S \cap B} h_p(pr_{T_pS}X_p, pr_{T_pS}X_p)dp \tag{1}$$

$$= \int_{S \cap B} q_p(pr_{T_pS}X_p)dp,$$

where $pr_{T_pS}$ denotes the orthogonal projection over the tangent plane $T_pS$ of $S$ at $p$. If $X$ is a constant vector fields in $\mathbb{E}^3$, for any fixed Borel subset, $\overline{q}_B(X)$ is a measure. $\{\lambda_{1_B}, \lambda_{2_B}, \lambda_{3_B}\}$ is the associated eigenvalue set, and $\{e_{1_B}, e_{2_B}, e_{3_B}\}$ is the eigenvector set of $\overline{h}_B$. The map $\lambda_i : B \to \lambda_{i_B}, i \in \{1,2,3\}$ is a measure of $\mathbb{E}^3$, named as *principal curvature measure*. Remark that, in piece-wise case, principal curvature measures have three components rather than two in the point-wise approach.

**Principal Curvature Measures of Triangle Meshes.** A triangle mesh is a discrete surface apparently, which means its shape and bending information can not be describe by point-wise approach. That's why the measure theoretic method is coherent in triangle mesh case. According to the concept and theory of normal cycle [4,27], suppose a triangle mesh $\mathcal{T}$ in $\mathbb{E}^3$, an explicit formula of $\overline{h}$ and $\overline{q}$ defined of constant vector field $X$ as:

$$\overline{h}_B(X,X) = \sum_{e \in E} l(e \cap B) \angle(e) <X,e><X,e>,$$

$$\overline{q}_B(X) = \sum_{e \in E} l(e \cap B) \angle(e) <X,e>^2. \tag{2}$$

where $E$ denotes the set of edge $e$ of $\mathcal{T}$, $l(e \cap B)$ denotes length of $e$ belongs to $B$, and $\angle e$ denotes the signed angle between unit normals $n_1$ and $n_2$ of incident facets $f_1$ and $f_2$ to $e$. Meanwhile, $\overline{h}_B$ associated matrix $F_B$ is written as:

$$F_B = \sum_{e \in E} l(e \cap B) \angle(e) e \cdot e^t. \tag{3}$$

We similarly name the set of eigenvalues $\{\lambda_{1_B}, \lambda_{2_B}, \lambda_{3_B}\}$ of $\overline{h}_B$ is the *principal curvature measures* of $\mathcal{T}$ over $B$. The corresponding set of eigenvectors $\{e_{1_B}, e_{2_B}, e_{3_B}\}$ of $\overline{h}_B$ can also be estimated. Based upon the generalization of

$\overline{h}_B$, three eigenvectors are respectively two principal directions and one normal direction of $X$ over $B$.

In summary, the set of principal curvature measures $\lambda_{1_B}, \lambda_{2_B}, \lambda_{3_B}$ is coherent to the geometrical properties of 3D face scans in triangle mesh and comprehensively suitable to describe facial surface. These principal curvature measures are the essential geometric attributes (*i.e.* the geometric feature) of our 3D keypoint descriptor presented in following parts.

## 3.2  $\lambda_B$ Based 3D Keypoint Detection

In order to guarantee the scale invariance property of facial descriptor, our keypoint detection is inspired by Lowe's SIFT [23] and related works [19,31], but the difference of principal curvature measures are used to locate keypoints. We firstly construct the Gaussian scale space by smoothing the face scan in triangle mesh with a series of Gaussian kernel $g_{\sigma_s}$ ($\sigma_s$ denotes to different scales). Given a vertex $v_i$ in a face scan, the facial surface adjacent becomes smoother by convolving $g_{\sigma_s}$ over neighbour vertices $v_j$ and $v_i$ is updated to $v_{i_{\sigma_s}}$ as (4).

$$v_{i_{\sigma_s}} = \frac{\sum_{v_j \in N(v_i,1)} g_{\sigma_s}(v_i, v_j) \cdot v_j}{\sum_{v_j \in N(v_i,1)} g_{\sigma_s}(v_i, v_j)} \tag{4}$$

where $N(v_i, 1)$ denotes the set of vertices within 1-ring neighbourhood of $v_i$ and Gaussian kernel $g_{\sigma_s}$ is defined as

$$g_{\sigma_s}(v_i, v_j) = exp(-\|v_i - v_j\|^2 / 2\sigma_s^2). \tag{5}$$

We estimate principal curvature measures $\lambda_{i_B}$ over each scale space of 3D facial surface as introduced in Sect. 3.1, and then compute the keypoint location criterion called Difference of Curvatutre (DoC), referring to Difference of Gaussian (DoG) in SIFT.

$$\delta(\lambda_i(B_{v_{\sigma_s}})) = \lambda_i(B_{v_{\sigma_s}}) - \lambda_i(B_{v_{\sigma_{s-1}}}), i = 1, 2, 3 \tag{6}$$

where $\delta$ denotes Difference of Curvature over $B$. If DoC associated to $v_i$ is the extreme among 1-ring vertices around $v_i$ in three scales $\sigma_{s-1}$, $\sigma_s$ and $\sigma_{s+1}$, $v_i$ is defined as a keypoint $v_k$ and $\sigma_s$ is its corresponding detection scale. The keypoints detected by three principal curvature measures separately are combined as one group for following 3D keypoint description.

## 3.3  $\lambda_B$ Based 3D Keypoint Description

3D keypoint description can be divided into two parts. The first one is to assign a primary direction for improving the robustness to minor head pose change. The second part is to construct histograms of curvature measures based feature descriptor.

**Primary Direction Determination.** Let's suppose one keypoint $v_k$ and its proper scale $\sigma_s$ of a face scan $F$ as before. The primary direction $\mathbf{d}_{v_k}$ associated to $v_k$ is defined by the neighbour vertices $\mathcal{N}(v_k)$ within a geodesic disc of $\sigma_s$ related radius $R_{\sigma_s}$:

$$\mathcal{N}(v_k) = \{v_j \in F | Dist(v_k, v_j) \leq R_{\sigma_s}\} \qquad (7)$$

We first determine a plane $TS_{v_k}$ orthogonal to the unit normal vector $\xi_{v_k}$ of $v_k$, then project the unit normal vector $\xi_{v_j}$ of $v_j$ on $TS_{v_k}$. The primary direction of $v_k$ is created by computing a Gaussian weighted histogram of 360 bins (1 bin per degree) on $TS_{v_k}$, and determined as the peak of the weighted direction histogram. Here the Gaussian weight is defined as:

$$w(v_k, v_j) = mag(v_j) \cdot g_{\sigma_s}(v_k, v_j),$$
$$mag(v_j) = \sqrt{\xi^x(v_j)^2 + \xi^y(v_j)^2}. \qquad (8)$$

**HoC Based Feature Descriptor Representation.** We construct the feature descriptor of a keypoint by using principal curvature measures estimated on a set of neighbor vertices. Following 2D daisy descriptor [37], all the neighbor vertices locating in 9 ovelapping circles $r_1, r_2, \cdots, r_9$ with a radius of $3.75\sigma_s$ around the keypoint support one keypoint feature descriptor. $r_1$ locates in the central part and its center is the keypoint here. Starting form the primary direction, other 8 circles around range along as clock-wise order (as shown in Fig. 1), and the distance from their center to the keypoint is $4.5\sigma_s$. This kind of daisy flower pattern descriptor simulates the functioning of human complex cells in visual cortex [10], and tends to be invariant to minor face transformation.

Then we build three histograms of three principal curvature measures ($hoc_i$) respectively in each circular region of $r_1, r_2, \cdots, r_9$. In each circular region, the value of $i^{th}$ principal curvature measure are quantized equally to 8 bins and weighted by Gaussian kernel, and the standard deviation is assumed as the Euclidean distance between current point to corresponding center of circle. After that, we normalize and concatenate all three principal curvature measures related 27 histograms (*3 principal curvature measures × 9 regions*) following this rule:

$$HOC = \{hoc_1^{r_1}, hoc_1^{r_2} \cdots hoc_1^{r_9}, hoc_2^{r_1}, hoc_2^{r_2} \cdots hoc_2^{r_9}, hoc_3^{r_1}, hoc_3^{r_2} \cdots hoc_3^{r_9}\} \qquad (9)$$

$HOC$, denoting to *Histogram Of principal Curvature measures*, is the keypoint feature descriptor.

### 3.4   3D Keypoint Matching

For keypoint matching, we aim to find matched keypoint pairs based on $HOC$ feature descriptor. Assume a keypoint $v_{k_i}^1$ belongs to first facial surface and the set of all keypoints $\{v_{k_j}^2\}$ in second facial surface. We estimate the angle set

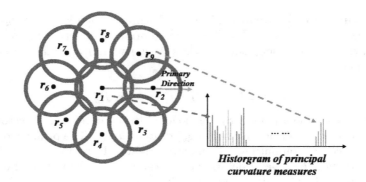

**Fig. 1.** Keypoint descriptor configuration in 9 overlapping circles for generating HoC.

$\{\alpha_j^i\}$ between feature vectors of $v_{k_i}^1$ and $\{v_{k_j}^2\}$ inspired to [31]. Each angle is defined as:

$$\alpha_j^i = cos^{-1}\left(\frac{<HOC_i^1, HOC_j^2>}{||HOC_i^1|| \cdot ||HOC_j^2||}\right). \tag{10}$$

The angles $\alpha_j^i$ are then ranked in ascending order. If the ratio between the first and second angle is smaller than predefined threshold $r_\alpha$, the match is accept. Otherwise it is rejected. Finally the number of matching keypoints is set as the similarity measurement $\mu$ between two facial surfaces.

## 4    Experiments

### 4.1    Database

In the experiment part, Morpho database and FRGC database are both involved for evaluation in different scenarios. We will introduce them briefly as follow.

**Morpho Database.** In Morpho databse, 16 masks were manufactured according to the facial information of 16 persons. Their faces are captured by 3D scanner with the structured light, and the mask is manufactured with 3D printer by Sculpteo 3D Printing [16]. Morpho database consists of two parts: (a) 20 subjects with 10 genuine face samples; (b) 20 subjects wearing their own or other's mask are captured around 10 times. In part (b), a person wearing his/her own mask is marked as a type $A_A$ mask sample. Otherwise it's marked as a type $A_B$ mask sample. In the following experiments, both $A_A$ and $A_B$ samples are both regarded as spoofing mask attacks. Some examples in Morpho database are shown in Fig. 2.

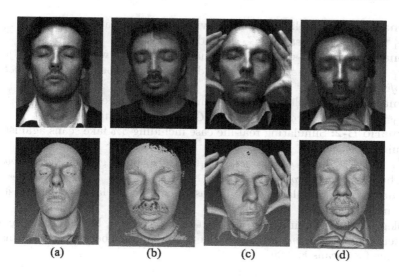

**Fig. 2.** 2D texture image and corresponding 3D mesh sample in Morpho database. (a) genuine face scan of person A; (b) genuine face scan of person B; (c) person A wears his own mask (type $A_A$ fake sample); (d) person A wears person B's mask (type $A_B$ fake sample).

**FRGC V2.0 Database.** FRGCv2 database is the largest published 3D face database, which is built with 4,007 3D face scans of 466 subjects with various facial expressions, genders and ages. All the face samples recorded are genuine faces. The face samples are captured in controlled pose and lighting condition by Minolta Vivid 900 scanner. After preprocessing step following [34], ROI contains about 30,000 vertices and 40,000 facets.

## 4.2 Experiment Scenarios

The basic purpose of our 3D face verification system is to guarantee the verification accuracy for genuine face, and the complementary purpose is to distinguish faces and masks. We hence firstly define a series of threshold of the similarity measurement $t_\mu^i$ in a baseline estimation scenario, and then we apply the same series of thresholds to evaluate the anti-spoofing performance of our facial feature descriptor. Furthermore, we also try to control the quantity ratio of genuine faces to fake faces in probe set for simulating a real-life case. Based on this idea, two series of experiment scenarios (Scenario A and B) are designed. For introducing clearly in following part, $G_M$ and $S_M$ denotes respectively to the set of all genuine face scans and the set of all spoofing mask scans (including type $A_A$ and $A_B$ mask scans) in Morpho database. $G_F$ denotes to the group of all genuine face scans in FRGC database. $G_{M1}$ and $G_{Mi}$ represents respectively the group of the first scan of individuals, and other samples of individuals in $G_M$ (similarly for $G_F$).

- **Scenario A-1:** Baseline evaluation with Morpho database.
  $G_{M1}$ forms *gallery set*, and $G_{Mi}$ builds *probe set*.
- **Scenario A-2:** Anti-spoofing performance evaluation with Morpho database.
  $G_{M1}$ forms *gallery set*, and $G_{Mi}$ and $S_M$ build respectively *probe set* and *spoofing probe set*.
- **Scenario B-1:** Baseline evaluation simulating real-life case.
  $G_{M1}$ and $G_{F1}$ build *gallery set*, and $G_{Mi}$ and $G_{Fi}$ build *probe set*.
- **Scenario B-2:** Simulating real-life case including 5% fake scans against 95% genuine face scans.
  $G_{M1}$ and $G_{F1}$ build *gallery set*, $G_{Mi}$ and $G_{Fi}$ build genuine face *probe set*. $S_M$ build *spoofing probe set*.
- **Scenario B-3:** Simulating real-life case including 1% fake scans against 99% genuine face scans.
  Gallery set and genuine face probe set is same as B-2, but only 20% scans of $S_M$ enrolled in *spoofing probe set*. The quantity ratio between fake scans and genuine face scans is 1:99.

### 4.3   Experimental Results and Analysis

In scenarios A-1 and B-1, we evaluate the baseline of verification performance with True Accept Rate (TAR). Because TAR varies along with $t_\mu^i$, TAR is shown in Table 1 under several False Accept Rate (FAR) cases. For scenarios A-2, B-2 and B-3 in Table 1, we first take only spoofing probe set to evaluate the anti-spoofing performance. Recall it that all scans in spoofing probe set are type $A_A$ or $A_B$ mask scans which are treated as illegal samples and should be rejected by system. True Accept Rate (TAR) and False Reject Rate (FRR), which are commonly used as criterion, can't be estimated in this case, because there are only two verification results for the spoofing probe sample, which are accepted falsely (FAR) or rejected correctly (TRR). Therefore we evaluate the performance with *Spoofing True Reject Rate* (STRR) as criterion for spoofing scans in scenarios A-2, B-2 and B-3 with FAR predefined in scenario A-1 and B-2 respectively. STRR is a special criteria for the spoofing samples, so as to show the distinction to TRR for the genuine face samples.

As shown in Table 1, the verification rate in baseline evaluation with Morpho Database is above 92% except the case that FAR is 0.01. And in real-life simulating case, we extend the scale of database by adding FRGC database and the verification rate is 91.98% even FAR equals only 0.001. If FAR is 0.01, the TAR increases from 84.75% to 94.68% which means the real-life simulating scenario with more samples can evaluate more properly the performance. Here, we can conclude that, in the baseline evaluation scenarios, our 3D face feature guarantees the verification performance for only genuine faces. Even though TAR is 84.75% in A-1, it can't deny the above conclusion. Because the scale of genuine face scans in A-1 is 180, which means there are only 1 or 2 samples accepted falsely if FAR is 0.01. And that's why we don't show the results of A-1 and A-2 when FAR is 0.001. By the way, based on the contrary thought, the results when FAR is 0.1 are blank in last three scenarios. Because there are 3,721 genuine face

**Table 1.** Verification and anti-spoofing performance evaluation in scenarios

| Scenarios | A-1 | A-2 | B-1 | B-2 | B-3 |
|---|---|---|---|---|---|
| FAR | TAR | STRR | TAR | STRR | STRR |
| 0.1 | 94.35% | 62.82% | - | - | - |
| 0.05 | 92.09% | 69.49% | 96.41% | 62.97% | 69.49% |
| 0.01 | 84.75% | 75.64% | 94.68% | 73.59% | 73.78% |
| 0.001 | - | - | 91.98% | 81.28% | 84.59% |

scans in scenario B-1, and if FAR is 0.1, there are too many samples accepted falsely which can't show the performance correctly.

In scenario A-2, STRR is above 62.82% and arrives 75.64% when FAR is 0.01. It's obvious that our algorithm can distinguish the genuine face and fraud mask from Morpho database. STRR arises along with decrease of FAR because threshold $t_\mu^i$ is more restricted. Moreover, in real-life simulating case, STRR is higher than 62.97% and achieves 81.28% and 84.59% including different spoofing probe set. In this case, STRR is lower a bit with same FAR than A-2 because the scale of gallery set is larger and $t_\mu^i$ is more rough. However, this experimental results also demonstrate our algorithm can complete the anti-spoofing mission even the mask manufactured in high quality.

After that, for scenarios A-2, B-2 and B-3, we combine genuine face related probe set ($G_{Mi}$ and $G_{Fi}$) and spoofing probe set ($S_M$) to evaluate the performance. Here a Detection Error Trade-off graph (DET) is given to show the experimental results (as shown in Fig. 3). Remark that in order to present clearly, the figure only shows the part when FAR and FRR is lower than 35% in scenarios B. In Fig. 3, the lower Equal Error Rate (ERR) is, with same gallery set, the better verification performance is. In this DET graph, when the scale of gallery set is limited in Morpho database, EER of A-2 is 9.3% higher than the baseline. It's obvious to study that the involvement of spoofing mask samples decreases the verification rate in scenarios A. In similar, when we extend the scale of gallery set enrolling FRGC database, EER of B-2 and B-3 is higher than corresponding baseline experiment. But comparing to A-2, EER declines to 3.1% and 2.8%. A similar conclusion obtained as before that two goals have been achieved: (1) the verification system guarantees a high verification ability and (2) it possesses the distinguishable power against spoofing attack in real-life simulating case.

## 4.4 Comparison with the State-of-the-Art Approaches

In this subsection the comparison with the state-of-the-art approaches using Morpho database is also given in Table 2. According to the experimental config-uration assigned in [7,17], here we use scenario A-1 to compute the EER. Then adopt the same threshold in A-2 to compute the SFAR, that is, *Spoofing False Accept Rate*. Spoofing False Accept indicates to the case that the samples with mask is false accepted by the sysem. In Table 2, we only report our experimental

**Table 2.** Comparison of verification performance with spoofing attacks in Morpho database. (1) Results reported in [17], (2) results reported in [7], (3) results with our proposed method.

| | Texture Image | | Depth Image | | 3D Mesh Model | | |
|---|---|---|---|---|---|---|---|
| | (1) | (2) | (1) | (2) | (1) | (2) | (3) |
| EER | 5.90% | 6.54% | 7.27% | 17.63% | 3.85% | 9.58% | 6.72% |
| SFAR | 72.87 | 59.94% | 88.94 | 47.98% | 91.46 | 54.09% | 33.10% |

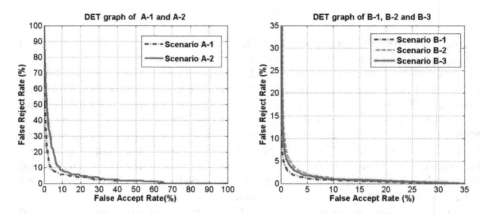

**Fig. 3.** Detection error trade-off graph of experimental scenarios: Scenarios A-1 and A-2 are shown in left graph, Scenarios B-1, B-2 and B-3 are shown in right graph.

results using 3D mesh samples. Comparing to 2D texture image and 2.5D depth image, the face recognition based on 3D face samples generally achieve higher verification performance with all genuine face samples in the test. The warping parameters related method in [17] achieves the lowest EER of 3.85%, which is better than 6.91% EER obtained by our method. However, the WP-related FR system is the most vulnerable one among the reported systems. Our PCM-meshSIFT-based method is the most robust system when replace all the probe samples by the samples with masks. We achieve the lowest SFAR of 33.10% in such experiment. It proves that the minor shape difference between genuine faces and manufactured masks can be detected and highlighted by our principal curvature measure based 3D facial feature, which is effective to enhance the security level of FR system.

## 5   Conclusion

In this paper, we first propose to using 3D shape description related method to distinguish the genuine faces and the spoofing masks stored in 3D triangle meshes. Due to the estimation process relying on a integral form, principal curvature measures are suitable to present the shape of triangle mesh directly.

Furthermore, principal curvature measures related feature descriptor can characterize properly the shape information of facial surface, and highlight the minor dissimilarity of shape between manufactured masks and genuine faces. Thereby our system can guarantee both high verification rate for genuine face and satisfactory anti-spoofing performance against mask attack.

Moreover, in a real-life case, the spoofing mask attack is a small proportion of testing samples. In experiment part, we hence propose to extend the probe set by combining mask samples in Morpho database and genuine faces in FRGCv2 database for simulating a real-life verification environment. The experimental results show that our method is effective in verification scenario and anti-spoofing performance during this simulating case.

**Acknowledgements.** This work was supported in part by the French research angency, l'Agence Nationale de Recherche (ANR), through the Biofence project under the grant **ANR-13-INSE-0004-02**.

# References

1. Bowyer, K.W., Chang, K., Flynn, P.: A survey of approaches and challenges in 3D and multi-modal 3D+2D face recognition, vol. 101, pp. 1–15. Elsevier (2006)
2. Bronstein, A.M., Bronstein, M.M., Kimmel, R.: Expression-invariant 3D face recognition. In: Kittler, J., Nixon, M.S. (eds.) AVBPA 2003. LNCS, vol. 2688, pp. 62–70. Springer, Heidelberg (2003). doi:10.1007/3-540-44887-X_8
3. Chetty, G., Wagner, M.: Multi-level liveness verification for face-voice biometric authentication. In: 2006 Biometrics Symposium: Special Session on Research at the Biometric Consortium Conference, pp. 1–6. IEEE (2006)
4. Cohen-Steiner, D., Morvan, J.M.: Restricted delaunay triangulations and normal cycle. In: ACM, pp. 312–321 (2003)
5. De Marsico, M., Nappi, M., Riccio, D., Dugelay, J.L.: Moving face spoofing detection via 3D projective invariants. In: 5th IAPR International Conference on Biometrics, pp. 73–78. IEEE (2012)
6. Erdogmus, N., Marcel, S.: Spoofing in 2D face recognition with 3D masks and anti-spoofing with kinect. In: IEEE International Conference on BTAS, pp. 1–6 (2013)
7. Erdogmus, N., Marcel, S.: Spoofing face recognition with 3D masks. IEEE Trans. Inf. Forensics Secur. **9**(7), 1084–1097 (2014)
8. Gordon, G.G.: Face recognition based on depth and curvature features. In: IEEE Computer Society Conference on CVPR, pp. 808–810 (1992)
9. Huang, D., Ardabilian, M., Wang, Y., Chen, L.: 3-D face recognition using elbp-based facial description and local feature hybrid matching. IEEE Trans. Inf. Forensics Secur. **7**(5), 1551–1565 (2012)
10. Hubel, D.H., Wiesel, T.N.: Receptive fields, binocular interaction and functional architecture in the cat's visual cortex. J. Physiol. **160**(1), 106–154 (1962)
11. Jain, A., Hong, L., Pankanti, S.: Biometric identification. Commun. ACM **43**(2), 90–98 (2000)
12. Jain, A.K., Ross, A., Prabhakar, S.: An introduction to biometric recognition. IEEE Trans. Circuits Syst. Video Technol. **14**(1), 4–20 (2004)

13. Kim, Y., Na, J., Yoon, S., Yi, J.: Masked fake face detection using radiance measurements. JOSA A **26**(4), 760–766 (2009)
14. Kollreider, K., Fronthaler, H., Bigun, J.: Evaluating liveness by face images and the structure tensor. In: IEEE Workshop on Automatic Identification Advanced Technologies, pp. 75–80 (2005)
15. Kollreider, K., Fronthaler, H., Bigun, J.: Verifying liveness by multiple experts in face biometrics. In: IEEE Computer Society Conference on CVPRW, pp. 1–6 (2008)
16. Kose, N., Dugelay, J.L.: Countermeasure for the protection of face recognition systems against mask attacks. In: 10th IEEE International Conference and Workshops on Automatic Face and Gesture Recognition (FG), pp. 1–6. IEEE (2013)
17. Kose, N., Dugelay, J.L.: On the vulnerability of face recognition systems to spoofing mask attacks. In: 2013 IEEE International Conference on Acoustics, Speech and Signal Processing, pp. 2357–2361. IEEE (2013)
18. Kose, N., Dugelay, J.L.: Shape and texture based countermeasure to protect face recognition systems against mask attacks. In: Proceedings of the IEEE Conference on Computer Vision and Pattern Recognition Workshops, pp. 111–116 (2013)
19. Li, H., Huang, D., Morvan, J.M., Wang, Y., Chen, L.: Towards 3D face recognition in the real: a registration-free approach using fine-grained matching of 3D keypoint descriptors. Int. J. Comput. Vision **113**(2), 128–142 (2015)
20. Li, J., Wang, Y., Tan, T., Jain, A.K.: Live face detection based on the analysis of fourier spectra. In: Defense and Security, pp. 296–303 (2004)
21. Li, X., Jia, T., Zhang, H.: Expression-insensitive 3D face recognition using sparse representation. In: IEEE Conference on Computer Vision and Pattern Recognition, pp. 2575–2582 (2009)
22. Li, X., Zhang, H.: Adapting geometric attributes for expression-invariant 3D face recognition. In: IEEE International Conference on Shape Modeling and Applications, pp. 21–32 (2007)
23. Lowe, D.G.: Distinctive image features from scale-invariant keypoints. Int. J. Comput. Vision **60**(2), 91–110 (2004)
24. Lu, X., Jain, A.K., Colbry, D.: Matching 2.5D face scans to 3D models. IEEE Trans. Pattern Anal. Mach. Intell. **28**(1), 31–43 (2006)
25. Määttä, J., Hadid, A., Pietikäinen, M.: Face spoofing detection from single images using micro-texture analysis. In: International Joint Conference on Biometrics, pp. 1–7. IEEE (2011)
26. Mian, A., Bennamoun, M., Owens, R.: An efficient multimodal 2D–3D hybrid approach to automatic face recognition. IEEE Trans. Pattern Anal. Mach. Intell. **29**(11), 1927–1943 (2007)
27. Morvan, J.M.: Generalized Curvatures. Springer, Heidelberg (2008)
28. Nixon, K.A., Aimale, V., Rowe, R.K.: Spoof detection schemes. In: Jain, A.K., Flynn, P., Ross, A.A. (eds.) Handbook of Biometrics, pp. 403–423. Springer, New York (2008)
29. Pan, G., Sun, L., Wu, Z., Lao, S.: Eyeblink-based anti-spoofing in face recognition from a generic webcamera. In: 11th IEEE International Conference on Computer Vision, pp. 1–8 (2007)
30. Pears, N., Liu, Y., Bunting, P. (eds.): 3D Imaging, Analysis and Applications, vol. 3. Springer, London (2012)
31. Smeets, D., Keustermans, J., Vandermeulen, D., Suetens, P.: meshSIFT: local surface features for 3D face recognition under expression variations and partial data. Comput. Vis. Image Underst. **117**(2), 158–169 (2013)

32. Sun, X., Morvan, J.M.: Curvature measures, normal cycles and asymptotic cones. Actes des rencontres du C.I.R.M. **3**(1), 3–10 (2013)
33. Sun, X., Morvan, J.M.: Asymptotic cones of embedded singular spaces. arXiv preprint arXiv:1501.02639 (2015)
34. Szeptycki, P., Ardabilian, M., Chen, L.: A coarse-to-fine curvature analysis-based rotation invariant 3D face landmarking. In: 3rd IEEE International Conference on Biometrics: Theory, Applications, and Systems, pp. 1–6 (2009)
35. Tanaka, H.T., Ikeda, M., Chiaki, H.: Curvature-based face surface recognition using spherical correlation. Principal directions for curved object recognition. In: 3rd IEEE International Conference on FG, pp. 372–377 (1998)
36. Tang, Y., Sun, X., Huang, D., Morvan, J.M., Wang, Y., Chen, L.: 3D face recognition with asymptotic cones based principal curvatures. In: IEEE International Conference on Biometrics, pp. 466–472 (2015)
37. Tola, E., Lepetit, V., Fua, P.: DAISY: an efficient dense descriptor applied to wide-baseline stereo. IEEE Trans. Pattern Anal. Mach. Intell. **32**(5), 815–830 (2010)
38. Wu, Z., Wang, Y., Pan, G.: 3D face recognition using local shape map. In: IEEE International Conference on Image Processing, vol. 3, pp. 2003–2006 (2004)
39. Zhang, Z., Yi, D., Lei, Z., Li, S.Z.: Face liveness detection by learning multispectral reflectance distributions. In: IEEE International Conference on Automatic Face & Gesture Recognition and Workshops, pp. 436–441 (2011)

# Early Features Fusion
# over 3D Face for Face Recognition

Claudio Tortorici and Naoufel Werghi$^{(\boxtimes)}$

Electrical and Computer Engineering,
Khalifa University, Abu Dhabi, UAE
{claudio.tortorici,naoufel.werghi}@kustar.ac.ae

**Abstract.** In this paper, a novel approach for fusing shape and texture Local Binary Patterns (LBP) for 3D Face Recognition is presented. Using the recently proposed *mesh-LBP* [23], it is now possible to compute LBP directly on a mesh manifold, allowing *Early Feature Fusion* to enhance face description power. Compared to its depth image counterparts, the proposed method (a) inherits the intrinsic advantages of mesh surfaces, (such as preservation of full geometry), (b) does not require face registration, (c) can accommodate partial or rotation matching, and (d) natively allows early-level fusion of texture and shape descriptors. The advantages of early-fusion is presented together with an experimentation of two merging schemes tested on the Bosphorus database.

**Keywords:** Local Binary Pattern · Early feature-fusion · LBP · 3D face recognition

## 1 Introduction

The last years have seen an extensive investigation of image usage for human identification and authentication. Even though biometric technologies, such as fingerprint and iris scan, seem to be more accurate, they require more human collaboration than face recognition techniques. Moreover, the creation of 3D imaging technologies has brought a further boost in the development of face recognition. In fact, the new generation of acquisition devices is now capable of capturing the geometry of 3D objects in the three-dimensional physical space.

Besides shape information, face imaging, in general, has emerged as promising modality with respect to other biometrics recognition techniques, such as universal acceptance and non-invasiveness. Moreover, 3D face imaging addresses some limitations of its 2D counterpart, like pose and luminance variation, while opening-up new horizons for enhancing the reliability of face-based identification systems [5]. This trend has been further fueled by the advances of 3D scanning technology, which provides now 3D textured scans encompassing aligned shape and photometric data.

In this paper, after a brief literature review, an explanation of Mesh-LBP framework is given (Sect. 2), then an outline of the proposed approach (Sect. 3.1), analyzing its potentialities; finally, some preliminary results are presented (Sect. 4) to support our proposal.

© Springer International Publishing AG 2017
B. Ben Amor et al. (Eds.): RFMI 2016, CCIS 684, pp. 56–64, 2017.
DOI: 10.1007/978-3-319-60654-5_5

## 1.1   Related Works

The state of the art is plenty of 3D face recognition approaches, making impossible to analyze all of them. Instead, we are going to present the works that guided our decisions categorizing them into three categories.

First we have approaches that base their strength in the local description given by *Fiducial Points*. Such methods use local representation of a face natively supporting partial matching, and in the last years are gaining credit in the community. In fact, a face can be described as a whole (*global representation*), or as combination of *local* partitions. Each partition, or region, is represented by a descriptor [25], and the combination of such descriptors is the representation of the face. Using fiducial points, it is also possible to get a face matching that handle face expressions distortions; it is in fact possible isolate and discard regions that are highly affected by such expression deformation, mouth and eyebrows above all. One of the first proposed approach [16], presented a *keypoints* detector based on SIFT [12]. However, it did not account partial scan and face rotation. Later on [14,19], presented a SIFT based method modeled to work on mesh manifold instead of standard flat images. That new born mesh-SIFT has been used in [9] together with the Sparse Representation based Classifier (SRC) [24] to boost the keypoint matching.

As second category is composed by all the Local Binary Pattern based approaches. LBP has been proposed in [2] as a 2D descriptor that well performed in texture retrieval problems. Given its success it has been applied to face recognition problem in [7], and later in 3D face recognition. In fact, LBP is now widely used on depth images [8,11], performing very well both from precision than performance perspectives. Moreover, LBP's versatility allowed building several variants. In [18] it has been introduced the Local Normal Binary Pattern (LNBP) that uses normals angle instead of depth values. 3D-LBP [20] works on a mesh computing the code using two kinds of values, one is the depth values and the other is the angle between normals of vertex of the mesh. Such approach however, requires an elaborated processing on the mesh in order to obtain the neighborhood of a central vertex. Moreover, 3D-LBP does not support multiple scale resolution like other previous LBP variants.

Finally, there is the group of Multi-modal 2D-3D approaches. Multi-modal solutions aim to combine different processing paths, usually 2D and 3D, into a single framework in order to overcome criticisms of individual approaches. In [6] Principal Component Analysis (PCA) is applied to depth images and standard images separately, then the outcomes are combined to get the final result. In [13] Iterative Closest Point (ICP) is used to register the 3D face model, and combined with Linear Discriminant Analysis (LDA) applied to the 2D image to avoid illumination and pose variation problems. Finally, [15] performs face registration, to avoid pose variations, region segmentation, to account local geometry changes, a filtering of the scans using SIFT and 3D Spherical Face Representation (SFR), and then a region wise matching with the remaining faces focusing on region robust to expression distortions.

## 2    Mesh-LBP

Our reference work generates Local Binary Patterns (LBP) over a real 3D support represented by triangular mesh manifold. In fact, LBP has been recently refiled in [21,23]. Since its definition [17] and its simplest application in face recognition [1], LBP is an 8-bit code obtained comparing pixels' values inside a 3 × 3 window; the outcome of this comparison can be 1 or 0, whether the difference with neighbors' values is grater or less than zero. This pattern can be extended at different scales by changing the windows dimension and adopting circular neighborhoods at different radii.

In [23], the LBP idea has been broadened to 2D-mesh manifolds implementing power and elegance of LBP on a real 3D support.

Instead of pixels, the mesh is composed by facets. In order to obtain an ordered ring around a generic central facet $f_c$, the algorithm searches adjacent facets $f_{out}$ and iteratively concatenate them as shown in Fig. 1. In such elegant way, it is now possible to generate a *ring-like* pattern at different radius scales. In fact, a new sequence of ordered $f_{out}$ facets on the ring outer corner can be extracted allowing the ring construction procedure to be iterated (as shown in Fig. 2), generating concentric rings around the initial central facet $f_c$.

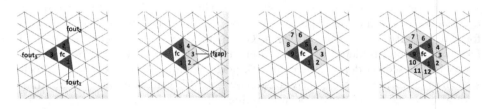

**Fig. 1.** From left to right the rings construction pipeline for mesh-LBP framework: starting from the adjacent facets, then the construction of the ring [22].

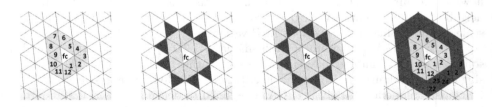

**Fig. 2.** Concentric rings construction sequence [22].

The concentric rings generated form an adequate structure for Local Binary Pattern computation. The mesh-LBP operator[1], around a generic central facet $f_c$, is defined as:

---

[1] The LBP descriptor complied with the mesh manifold.

$$meshLBP_m^r(f_c) = \sum_{k=0}^{m-1} s\left(h\left(f_k^r\right) - h\left(f_c\right)\right) \cdot \alpha(k) , \tag{1}$$

$$\text{with} \quad s(x) = \begin{cases} 1 \ x \geq 0 \\ 0 \ x < 0 \end{cases}$$

where parameters $r$ and $m$ control respectively the radial resolution and the azimuth quantization (see Fig. 2). Furthermore, a function $\alpha(k)$ has been introduced to derive different LBP variants. In this work two variant have been studied:

- $\alpha_2(k) = 2^k$, as originally suggested in [17];
- $\alpha_1(k) = 1$, to obtain a simplified form that sum the binary pattern digits.

In Sect. 4 we will refer to these two function with $\alpha_2$ and $\alpha_1$ respectively. $h(f)$ function can be any desired feature; it can represent shape or appearance information, depending on the feature used. For example, as shape descriptor a geometric feature can be extracted from the mesh surface, such as *mean curvature* or *curvedness*, rather than *gray level* values to represent appearance information. Such photometric values come from 2D flat images, acquired with standard cameras, and subsequently projected over the mesh using a mapping scheme embedded in the mesh itself.

## 3  Fusion Schemes

In order to proceed, a brief description of Face Recognition pipeline has to be presented. Mesh-LBP framework presented in [23] can be summarized in 5 main steps:

**Features extraction,** since a mesh manifold is a structure, some features have to be extracted in order to describe the shape of the mesh surface.

**Local Binary Pattern computation,** applying Eq. 1 using the features beforehand extracted as input data.

**3D grid construction,** a grid is constructed and projected on the mesh manifold focusing on some stable region of the face.

**Histograms computation and concatenation,** for each point of the grid, a region is defined and an histogram computed inside it; the concatenation of all the region histograms form a signature for the examined face scan.

**Face matching,** checks differences between probe scan and a defined gallery.

As this framework operates at different level over the same structure, it is possible to perform descriptors fusion at each level of the pipeline. In [22] has been shown how a simple score fusion, between geometric and photometric descriptors, fits or sometimes even outperforms the state of the art [4,10]. Furthermore, it presents two fusion schemes at histograms computation level: one concatenates two different histograms derived from geometric and photometric features (*region histograms concatenation*); while the other one counts the co-occurrences of the two features (*2D-histogram*). Such fusions show the potentiality of climbing the face matching pipeline to merge different descriptors.

The idea proposed in this paper is to do a step forward and make the fusion at Mesh-LBP computation level. Even if the results displayed in [22] show high accuracy rate, the histograms fusion introduces an increment of the face descriptor size. In fact, the more simple *region histograms concatenation* doubles the original histogram size, while *2D histogram*, that adds one dimension to the standard histograms, sees a geometric increment of size. Instead, if the fusion is performed during, or even before, the mesh-LBP computation, it is possible to use both geometric and photometric data, keeping dimension and size equal to a single descriptor. Our aim is to produce a descriptor that holds the same size obtained with a single feature, but the information of two features (shape and appearance in our case[2]).

### 3.1   Early-Fusion

In this paper two kinds of early-fusion are presented. The first is a very basic fusion scheme that use logic operators (*AND*, *OR* and *XOR*). In order to get the LBP code, such operators have been added to the original formula:

$$meshLBP = \begin{cases} AND(s_g(x), s_p(x)) \\ OR(s_g(x), s_p(x)) \\ XOR(s_g(x), s_p(x)) \end{cases} \qquad (2)$$

where $s_g(x)$ and $s_p(x)$ are computed as $s(x)$ in Eq. 1 respectively for geometric and photometric information.

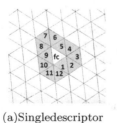

(a)Singledescriptor                    (b)Interleavingscheme

**Fig. 3.** Graphic comparison between standard ordering of a ring with single descriptor and the interleaving scheme with two descriptors.

In the second variant the mesh-LBP pattern is generated replacing the single feature function $h(f)$, shown in Eq. 1, with a combination of extracted features $h_g(f)$ and $h_p(f)$. In particular, such new descriptor, named $h_{g,p}(f)$, is composed by interleaved values from geometric and photometric data,

---

[2] The framework allows to use any kind of features.

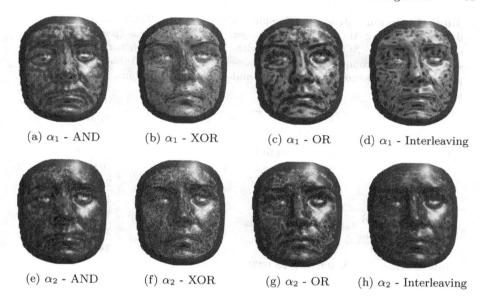

(a) $\alpha_1$ - AND        (b) $\alpha_1$ - XOR        (c) $\alpha_1$ - OR        (d) $\alpha_1$ - Interleaving

(e) $\alpha_2$ - AND        (f) $\alpha_2$ - XOR        (g) $\alpha_2$ - OR        (h) $\alpha_2$ - Interleaving

**Fig. 4.** Visual representation of *early-fusion* mesh-LBP code generated with radius $r = 4$ and azimuth quantization $m = 12$ in both $\alpha_1$ and $\alpha_2$ variants.

respectively $d^g$ and $d^p$ (Fig. 3). For example, for an azimuth quantization $m = 12$, the $h_{g,p}(f)$ sequence would be

$$h_{g,p}(f) = d_1^g, d_2^p, d_3^g, d_4^p, d_5^g, d_6^p, d_7^g, d_8^p, d_9^g, d_{10}^p, d_{11}^g, d_{12}^p \tag{3}$$

Successively, the mesh-LBP code is obtained from the new combination $h_{g,p}(f)$ applying Eq. 1 (Fig. 4).

From now on, these two variants will be referred to as *Logic Fusion* AND/OR/XOR and *Interleaving Fusion* respectively.

## 4    Experimentation

Experiments have been conducted on Bosphorus database [3], that is composed by 4666 scans of 105 subjects scanned in different poses, action units, and occlusion conditions. In addition to the shape structure, represented as mesh manifold, the database contains bitmap images of the scanned subject to provide appearance information as well. Since the aim of the project is to build a new LBP-like descriptor that can embed the strong points of a 3D environment, we did not focus on the matching algorithm. A naive template-matching-like method has been used, where each face probe descriptor is compared with a reference gallery using $\chi^2$ distance.

Comparing our results with [22], the same features have been chosen to be merged. In particular in Table 1 we show results obtained using the *mean curvature* to represent shape information, and the *graylevel*, got from the bitmap mapped on the mesh surface, for the appearance.

**Table 1.** Overall outcomes of Bosphorus database showing the accuracy of *Logic* operators (AND, XOR, OR) and *Interleaving scheme* compared with [22] single descriptors *mean curvature* (H) and *graylevel* (GL).

| | Accuracy | |
|---|---|---|
| | $\alpha_1$ | $\alpha_2$ |
| H | 86.82% | 90.70% |
| GL | 90.89% | 90.26% |
| Fusion$_1$ H+GL | 93.84% | – |
| Fusion$_2$ H+GL | 93.29% | 92.60% |
| Logic AND H+GL | 86.24% | 89.14% |
| Logic XOR H+GL | 77.30% | 90.35% |
| Logic OR H+GL | 85.99% | 89.79% |
| **Interleaving H+GL** | **92.76%** | **93.03%** |

**Table 2.** Histogram sizes (per region) for each variant reported in Table 1 in number of bins.

| Hist size | |
|---|---|
| $\alpha_1$ | $\alpha_2$ |
| 13 | 1125 |
| 13 | 136 |
| 169 | – |
| 26 | 1261 |
| 13 | 1125 |
| 13 | 1125 |
| 13 | 1125 |
| **13** | **1125** |

Results from logic fusions show an accuracy rate close to the original single descriptor. Even if the size of logic descriptor is equal to a single one, the outcomes are not satisfying: this scheme shows a decrease in its descriptive power respect to what has been achieved in our reference paper. In fact, logic operators seem to annihilate the mutual information provided by the couple of features.

Interleaving scheme, instead, preserves the descriptive power of both geometric and photometric information, outperforming single descriptor precision and above mentioned histograms fusions. In particular, $\alpha_1$, even if a bit lower in accuracy compared with *Fusion$_1$* and *Fusion$_2$* schemes, sees a drastic decrease of descriptor size (Table 2): half respect to *region histograms concatenation* schema, and even of the order of root square respect to the *2D-histogram* (13 times smaller). $\alpha_2$, instead, does not only keep the same size of single feature histogram, but also outperforms the region histograms concatenation fusion scheme.

The effectiveness of *Interleaving* early-fusion approach become clear if we think that 2D-histogram fusion scheme, cannot be computed for $\alpha_2$, that is the original LBP variant. In that case, in fact, the 2D-histogram would have had $1125 \times 136 = 153000$ bins instead of the 1125 of our proposed fusion scheme.

## 5 Conclusion

In this paper a novel early level fusion approach for actual 3D face recognition has been presented. The proposed method exploits mesh manifold potentialities as support structure. In particular, we extended mesh-LBP, a framework that enables to generate LBP-like codes directly on a triangular mesh. Our aim is to fuse different features during, or even before, the LBP descriptor computation. For this purpose logic operators and interleaving schemes have been used to generate a pattern comprehensive of photometric texture and geometric shape information. The experimentation, conducted on Bosphorus database, shows promising results, raising the curtains on the potentiality held by early feature fusion among real 3D support, like mesh manifolds. It is in fact now possible to consider more refined early-fusion techniques directly employed on a mesh manifold. In this manner, we can hold the descriptive power of two, or even more, descriptor, improving performances without increasing the descriptor size.

# References

1. Ahonen, T., Hadid, A., Pietikäinen, M.: Face recognition with local binary patterns. In: European Conference on Computer Vision, Prague, pp. 469–481, May 2004
2. Ahonen, T., Hadid, A., Pietikäinen, M.: Face recognition with local binary patterns. In: Pajdla, T., Matas, J. (eds.) ECCV 2004. LNCS, vol. 3021, pp. 469–481. Springer, Heidelberg (2004). doi:10.1007/978-3-540-24670-1_36
3. Alyüz, N., Gökberk, B., Akarun, L.: 3D face recognition system for expression and occlusion invariance. In: IEEE International Conference on Biometrics: Theory, Applications, and Systems, Washington, DC, pp. 1–7, September 2008
4. Berretti, S., Werghi, N., Del Bimbo, A., Pala, P.: Matching 3D face scans using interest points and local histogram descriptors. Comput. Graph. **37**(5), 509–525 (2013)
5. Bowyer, K.W., Chang, K.I., Flynn, P.J.: A survey of approaches and challenges in 3D and multi-modal 3D+2D face recognition. Comput. Vis. Image Underst. **101**(1), 1–15 (2006)
6. Chang, K., Bowyer, K., Flynn, P.: An evaluation of multimodal 2-D and 3-D face biometrics. IEEE Trans. Pattern Anal. Mach. Intell. **27**(4), 619–624 (2005)
7. Huang, D., Shan, C., Ardabilian, M., Wang, Y., Chen, L.: Local binary patterns and its application to facial image analysis: a survey. IEEE Trans. Syst. Man Cybern. Part C Appl. Rev. **41**(6), 765–781 (2011)
8. Huang, Y., Wang, Y., Tan, T.: Combining statistics of geometrical and correlative features for 3D face recognition. In: British Machine Vision Conference, Edinburgh, pp. 879–888, September 2006
9. Li, H., Chen, L., Huang, D., Wang, Y., Morvan, J.: Towards 3D face recognition in the real: a registration-free approach using fine-grained matching of 3D keypoint descriptors. Int. J. Comput. Vis. **113**(2), 128–142 (2015)
10. Li, H., Huang, D., Lemaire, P., Morvan, J.M., Chen, L.: Expression robust 3D face recognition via mesh-based histograms of multiple order surface differential quantities. In: IEEE International Conference on Image Processing, pp. 3053–3056, September 2011
11. Li, S., Zhao, C., Ao, M., Lei, Z.: Learning to fuse 3D+2D based face recognition at both feature and decision levels. In: International Workshop on Analysis and Modeling of Faces and Gestures, Beijing, pp. 44–54, October 2005
12. Lowe, D.G.: Distinctive image features from scale-invariant keypoints. Int. J. Comput. Vis. **60**(2), 91–110 (2004)
13. Lu, X., Jain, A.K.: Deformation modeling for robust 3D face matching. In: IEEE International Conference on Computer Vision and Pattern Recognition, New York, pp. 1377–1383, June 2006

14. Maes, C., Fabry, T., Keustermans, J., Smeets, D., Suetens, P., Vandermeulen, D.: Feature detection on 3D face surfaces for pose normalisation and recognition. In: 2010 Fourth IEEE International Conference on Biometrics: Theory Applications and Systems (BTAS), pp. 1–6. IEEE (2010)
15. Mian, A.S., Bennamoun, M., Owens, R.: An efficient multimodal 2D-3D hybrid approach to automatic face recognition. IEEE Trans. Pattern Anal. Mach. Intell. **29**(11), 1927–1943 (2007)
16. Mian, A.S., Bennamoun, M., Owens, R.: Keypoint detection and local feature matching for textured 3D face recognition. Int. J. Comput. Vis. **79**(1), 1–12 (2008)
17. Ojala, T., Pietikäinen, M., Harwood, D.: A comparative study of texture measures with classification based on featured distribution. Pattern Recognit. **29**(1), 51–59 (1996)
18. Sandbach, G., Zafeiriou, S., Pantic, M.: Local normal binary patterns for 3D facial action unit detection. In: IEEE International Conference on Image Processing, Orlando, pp. 1813–1816, September 2012
19. Smeets, D., Keustermans, J., Vandermeulen, D., Suetens, P.: meshSIFT: local surface features for 3D face recognition under expression variations and partial data. Comput. Vis. Image Underst. **117**(2), 158–169 (2013)
20. Tang, H., Yin, B., Sun, Y., Hu, Y.: 3D face recognition using local binary patterns. Signal Process. **93**(8), 2190–2198 (2013)
21. Werghi, N., Berretti, S., Del Bimbo, A., Pala, P.: The mesh-LBP: computing local binary patterns on discrete manifolds. In: ICCV International Workshop on 3D Representation and Recognition, Sydney, pp. 562–569, December 2013
22. Werghi, N., Tortorici, C., Berretti, S., Del Bimbo, A.: Representing 3D texture on mesh manifolds for retrieval and recognition applications. In: IEEE Conference on Computer Vision and Pattern Recognition (CVPR), Boston, pp. 2521–2530, June 2015
23. Werghi, N., Tortorici, C., Berretti, S., del Bimbo, A.: Local binary patterns on triangular meshes: concept and applications. Comput. Vis. Image Underst. **139**, 161–177 (2015)
24. Wright, J., Yang, A., Ganesh, A., Sastry, S., Ma, Y.: Robust face recognition via sparse representation. IEEE Trans. Pattern Anal. Mach. Intell. **31**(2), 210–227 (2009)
25. Zhao, W., Chellappa, R., Phillips, P.J., Rosenfeld, A.: Face recognition: a literature survey. ACM Comput. Surv. (CSUR) **35**(4), 399–458 (2003)

# Enhancing 3D Face Recognition by a Robust Version of ICP Based on the Three Polar Representation

Amal Rihani$^{(\boxtimes)}$, Majdi Jribi$^{(\boxtimes)}$, and Faouzi Ghorbel$^{(\boxtimes)}$

CRISTAL Laboratory, GRIFT Research Group,
National School of Computer Science,
University of Manouba, 2010 La Manouba, Tunisia
`amal.rihani@ensi-uma.tn`,
{`majdi.jribi,faouzi.ghorbel`}`@ensi.rnu.tn`

**Abstract.** In this paper, we intend to propose a framework for the description and the matching of three dimensional faces. Our starting point is the representation of the 3D face by an invariant description under the $M(3)$ group of translations and rotations. This representation is materialized by the points of the arc-length reparametrization of all the level curves of the three polar representation. These points are indexed by their level curve number and their position in each level. With this type of description we need a step of registration to align 3D faces with different expressions. Therefore, we propose to use a robust version of the iterative closest point algorithm (ICP) adopted to 3D face recognition context. We test the accuracy of our approach on a part of the BU-3DFE database of 3D faces. The obtained results for many protocols of the identification scenario show the performance of such framework.

**Keywords:** 3D face · Description · Three polar representation · The arc-length reparametrization · Registration · Haussdorff · ICP · BU-3DFE

## 1 Introduction

The automatic 3D shape recognition has known a growing interest during the last years in the pattern recognition field. Recently, the 3D data become active especially with the 3D acquisition materials improvement and the big computer capacity in the term of calculations. Therefore, the quality and the resolution of 3D meshes become better. In addition, 3D data permit to overcome the problems often encountered in 2D data. In fact, 2D data need an invariance under the perspective transformations while the 3D data surfaces need only the invariance under the Euclidian transformations. But one of the major problems of 3D surfaces is the lack of a canonical parameterizations. This fact makes hard the matching procedure between 3D objects. In order to overcome as much as possible this limits many works propose to extract an invariant description from

© Springer International Publishing AG 2017
B. Ben Amor et al. (Eds.): RFMI 2016, CCIS 684, pp. 65–74, 2017.
DOI: 10.1007/978-3-319-60654-5_6

3D surfaces under the initial parametrization. In the literature, the 3D shape description can be classified into two main categories: The global methods and the local ones.

Several 3D global surface descriptions were proposed in the literature. In this category, we can mention the cords histogram methods proposed by Paquet et al. [1]. Its consists on the extraction of the statistical characteristics from the cords of the 3D object. Osada et al. [2] proposed as a global description for the 3D surfaces, the 3D distribution forms method. This last one is obtained by a probability distribution of a 3D shape function.

For the second category of methods, a 3D local representation is extracted from a 3D objects. In this context, there are many local descriptors based on the curvature such as the Gaussian curvature proposed by Shw-wei et al. [3] which is used to describe the 3D faces. Also, Ganguly et al. [4] proposed to use a two pairwise of curvature analysis. The first pair is composed by the mean, and the maximum curvature and the second one corresponds to the minimum and the gaussian curvature. We can mention here, Bannour et al. [19] who presented a 3D surface description by a set of invariant points obtained from a set of uniform levels of the curvature values. Another kind of the local methods which based on the construction of the geodesic level curves around a feature point are used to represent the 3D surfaces. [5–7] proposed to describe the 3D surface by a set of geodesic level curves generated from a one reference point qualified by the unipolar representation. Other works proposed to use the representation based on many reference points in order to overcome the problem of the instability in the case of error of the reference point extraction. Ghorbel et al. [8] proposed to use the bipolar representation. It is obtained from two reference points. It consists on the levels of the superposition of the two geodesic potentials generated from two reference points. In this context, Jribi et al. [9] proposed to extend this representation to the three polar one based on the superposition of three geodesic potentials from three reference points instead of two.

The majority of these description methods require a registration step in order to estimate the variation between two shapes and to align them. In the literature, the registration methods between 3D shapes can be classified into two major categories. The first type is based on the local geometry to construct a valid hypotheses of mappings. In this category, we can classify the registration methods based on Hough transform and Hashage tables [15–18]. The second type performs the mapping by iterative algorithms. We can mention here the works of Bes et al. [12] who used an iterative techniques to extract the matched points. In this paper, we intend to propose a 3D face recognition technique based on two stages: The first one consists on the proposition of an invariant 3D face description. The second stage is a step of alignment of the 3D surface by a novel robust version of ICP [12].

Thus, this paper will structured as follows: we present in the second section a brief recall of the proposed representation. The implementation steps of the proposed representation on 3D faces are described in section three. The used similarity metric to compare between two shapes and the novel robust version of ICP are detailed in the fourth section and finally, we test the accuracy of our representation for the identification scenario on a part of the BU-3DFE database of 3D faces in the last section.

## 2 Brief Recall of the Proposed Representation

In this paper, we propose to describe the 3D surfaces by an accurate, finite, and invariant set of points under the geometrical transformations of the M(3) group of translations and rotations. This description is proposed by Rihani et al. [14]. It is obtained by two steps: (i) The first step consist on the construction of the three polar representation proposed by Jribi et al. [9]. (ii) in the second step a geometric arc-length reparametrization of each level of the three polar representation should be performed. We describe in the rest of the section the two steps cited above.

In the rest of the section, we consider that a 3D object as a 2D-differential manifold denoted by $S$.

### 2.1   The Construction of the Three Polar Representation

Let denote by $U_r$ the function that computes for each point $p$ of $S$ the length of the geodesic curve joining it to the point $r$. The three polar representation consists on the superposition of three geodesic potential generated from three reference points. Therefore let denote by $p_1$, $p_2$, $p_3$ three reference points of $S$ $U_{p_1}$, $U_{p_2}$, $U_{p_3}$ their corresponding geodesic potentials and $U_s$ the sum of these three geodesic potentials. Thus, the three polar representation that we denote by $M^k(S)$ corresponds to the set of $k$ level curves where each level curve $C^{\lambda_i}$ is composed by a set of points having the sum of the three geodesic potential $U_s$ equal to $\lambda$. It can be formulated as follows:

$$M^k(S) = \{C^{\lambda_i}\}_{i=1..k} \tag{1}$$

where

$$C^{\lambda_i} = \{p \in S, U_s(p) = \lambda_i\} \tag{2}$$

### 2.2   Geometric Arc-Length Reparametrization

After the construction of the three polar representation, the 3D surface $S$ is presented by a collection of level curves $\{C^{\lambda_i}\}$. A curve parametrization $\{C^{\lambda_i}(t)\}$ is an 1-periodic function of a continuous parameter $t$ defined by:

$$C^{\lambda_i} : [0, 1] \to \mathbb{R}^3 \tag{3}$$
$$t \mapsto [x(t), y(t), z(t)]^t$$

It's well known that the same parametric curve $C^{\lambda_i}$ can have many parameterizations. This due to parametrization dependance on the position, the orientation of the used curve and the speed we go over. In order to overcome this problem, we propose to use a $\mathbb{G}$ invariant reparametrization of each curve of the three polar representation where $\mathbb{G}$ is a group of geometrical transformations applied to a curve.

In our context, $\mathbb{G}$ corresponds to the $M(3)$ group formed by the $\mathbb{R}^3$ rotations and translations. This group of transformations preserves the length of the curve however the speed we go over the curve affects its parametrization. Therefore, we carry out an arc-length reparametrization of a 3D curve $C^{\lambda_i}$ in order to cover it with the same speed. The arc-length reparametrization is defined as follows:

$$S(t) = 1/L \int_0^t \sqrt{x(t)'^2 + y(t)'^2 + z(t)'^2} dt, t \in [0, T] \tag{4}$$

Here, $L$ denotes the length of the level curve $C^{\lambda_i}$.

## 3   The Application of the Proposed Representation on 3D Faces Meshes

Since the 3D faces known a growing interest for the identities determination especially after the many terrorist acts occurred around the world, we implement this novel representation on this type of data. In practice, the 3D surface corresponds to a discrete mesh. We will start by the construction of the three polar representation on the 3D faces. As mentioned before the three polar representation is based on the three reference points. In our case, the out corner of the eyes and the noise tip are used as reference points. For the automatic extraction of the reference points, we use an approach based on a curvature analysis of 3D faces proposed by Szeptycki et al. [21]. Then, for each reference point we compute its geodesic potential. In the discrete case, the computation of a geodesic potential generated from a reference point corresponds to the computation of the geodesic curves between the reference point and the other points of the 3D face. Here, we use the fast marching algorithm [13] for the computation of the geodesic path between each pairs of points. The three polar representation is composed by a set of discrete level curves. Each level curve of value $\lambda$ can be represented by a set of vertices. The sum of three geodesic potentials of each vertex should belongs to $[\lambda - \epsilon, \lambda + \epsilon]$ it can formulated as follows:

$$C^\lambda = \{P \in S, \lambda - \epsilon \le U_3(P) \le \lambda + \epsilon\} \tag{5}$$

where $\epsilon$ is a real positive value chosen according to the resolution of mesh to avoid the intersections between successive level curves.

After the construction of the geodesic level curves of the three polar representation, we perform the approximation of these curves by the B-spline function. Finally, we realize the arc-length reparametrization procedure for each level curve of the tree polar representation. The obtained points are equidistant and invariant under the $M(3)$ group of translations and rotations. Each point is defined by its level number value and its position in that level. In fact, the 3D face can be defined for $N$ levels of the three polar level curve by:

$$\hat{M}^N(S) = \{p_{ij}^S \in \mathbb{R}^3\}, i \in [1..N], j \in [1..L] \tag{6}$$

where $N$ is number of the geodesic level curves of the three polar representation and $L$ is number the points by level. In Fig. 1, we summarize all the steps of the proposed representation applied on the 3D faces.

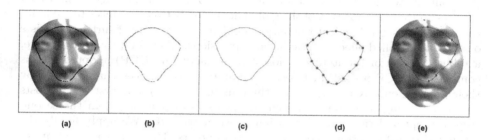

(a)          (b)          (c)          (d)          (e)

**Fig. 1.** The steps of the proposed approach applied to a 3D face. (a, b): The extraction of the three polar level curve. (c): Approximation of this level curve with the B-spline function. (d, e): The arc-length reparametrization of this level curve.

## 4  3D Faces Comparison

### 4.1  Haussdorff Shape Distance

In this work, we use the well known Haussdorff shape distance introduced by Ghorbel et al. [10, 11] for the recognition task between 3D shapes. All the possible parameterizations of surface are grouped on $G$. $G$ can be $\mathbb{R}^2$ plane if the surface is open or $\mathbb{S}^2$ if it is closed. Let $S_1$ and $S_2$ be two 3D surface pieces diffeomorphic to $G$ on which act the $M(3)$ group of geometrical transformations. The Hausdorff shape distance between $S_1$ and $S_2$ can be defined by:

$$\triangle(S_1, S_2) = max(\rho(S_1, S_2), \rho(S_2, S_1)) \tag{7}$$

where :

$$\rho(S_1, S_2) = \sup_{g_1 \in M(3)} \inf_{g_2 \in M(3)} \|g_1 S_1 - g_2 S_2\|_{L^2}^2 \tag{8}$$

Since the $M(3)$ displacement group preserves this norm, the Hausdorff shape distance can be written as the following quantity:

$$\triangle(S_1, S_2) = \inf_{h \in M(3)} \|S_1 - hS_2\|_{L^2}^2 \tag{9}$$

The transformation between two shapes should be estimated in order to compute the correct value of the Haussdorff shape distance. We use in our context, a novel robust version of the Iterative Closest Point algorithm to estimate the optimal transformation between faces. In this work, each 3D face is characterized by its 3D descriptor. Therefore a face is described by a set of infinite points obtained after the reparametrization of the tree-polar geodesic level curves.

### 4.2   Proposed Robust Version of ICP

In this work, we are interested on the problem of the 3D faces recognition. In this context, we generally need an elementary process of fine alignment which consists on the minimization of the global deviation between surfaces to compute the right distance value. But the major problem of such type of surfaces consists on the uncontrolled effects of the facial expressions. Therefore, we propose here, a robust version of the iterative closest point algorithm (ICP) adopted to this context. The ICP algorithm takes as input two 3D surfaces characterized by their points cloud. ICP is based on three main steps: $(i)$ The first one consists on the matching procedure between the two sets of points. $(ii)$ In the second step, the optimal rigid transformation is estimated. $(iii)$ We apply finally the estimated transformation to one of the sets of points. The main contributions of the proposed version of ICP are essentially in its two first steps.

Here, a 3D face is represented by a set of discrete points corresponding to the proposed descriptor. The descriptor of a 3D face $S_1$ is formulated as mentioned above by:

$$\hat{M}^N(S_1) = \{p_{ij}^{S_1} \in \mathbb{R}^3\}, i \in [1..N], j \in [1..L] \tag{10}$$

where $N$ is number of the three polar level curve of the three polar representation and $L$ is number the points by level.

Let consider two surfaces $S_1$ and $S_2$ and their respectively corresponding descriptors $\hat{M}^N(S_1)$ and $\hat{M}^N(S_2)$ are defined by:

$$\hat{M}^N(S_1) = \{p_{ij}^{S_1} \in \mathbb{R}^3\} \tag{11}$$
$$\hat{M}^N(S_2) = \{p_{ij}^{S_2} \in \mathbb{R}^3\}, i \in [1..N], j \in [1..L]$$

**First Step: Pairwise Points Matching.** Bes et al. [12] determined that the matching step assumed 95% of the ICP's time. This fact shows that the efficiency of the ICP depends on the corresponding step. In our approach, the 3D surface is presented by a set of discrete points. These points are indexed by their level number value and their position in this level. The first contribution of the proposed robust version of the ICP derive directly from the three polar representation. In fact, the matching procedure is automatically obtained since each point $p_{ij}^{S_1}$ is matched to the point $p_{ij}^{S_2}$ of the second face. One the other hand, a correct correspondence is conditioned by having a unique way to obtain the starting point on each level curve. We use, therefore, the plane passing through the noise tip and the first level of the three polar representation (which correspondence to a invariant point) to detect the starting point in each level curve.

The intersection between this plane and the 3D surface in each level curve of the three polar representation corresponds to the starting points of each three polar level curve.

**Second Step: Transformation Estimation.** The second step of ICP consists on the estimation of the rigid transformation between $\hat{M}^N(S_1)$ and $\hat{M}^N(S_2)$ that we denote by $\hat{T}$. ICP algorithm is an iterative procedure minimizing the Mean Square Error (MSE). In practice, the rigid transformation should find a solution to the least squares defined by:

$$\hat{T} = \operatorname*{argmin}_{T} \sum_i \sum_j e_{ij}^2 \tag{12}$$

where $e_{ij}$ is the distance between the point $p_{ij}^{S_1}$ of $S_1$ and its corresponding point $p_{ij}^{S_2}$ of $S_2$.

$$e_{ij}^2 = \|p_{ij}^{S_2} - T(p_{ij}^{S_1})\|^2 \tag{13}$$

Our approach is implemented on the 3D faces with different facial expressions. Since the rigid matching process is sensitive to the 3D shape deformations, we should consider this variation shape. In the present work, we propose to automatically associate different weights to the different points representing the 3D surface. In fact, only the points that are less influenced by the facial expressions will participate in this estimation step. To distinguish these points, we suggest to study the variation $V_{ij}^k$ of each point $p_{ij}^{S_k}$ of the surface $S_k$ from its centroid noted by $C_{S_k}$ in all the surfaces. This variation corresponds to the distances between $p_{ij}^{S_k}$ and $C_{S_k}$. It can defined by:

$$V_{ij}^k = (d(P_{ij}^k, C_{S_k})), \tag{14}$$

The weight value $W_{ij}^{S_k}$ given for each point $p_{ij}^{S_k}$ should qualify the quality of matching. Indeed, The more static the point is, the greater its weight should. Therefore, the weight $W_{ij}$ for two corresponding points $p_{ij}^{S_1}$ and $p_{ij}^{S_2}$ for the two surfaces $S_1$ and $S_2$ can be formulated by:

$$W_{ij} = \frac{V_{max} - (V_{ij}^{S_2} - V_{ij}^{S_1})}{V_{max}} \tag{15}$$

where $V_{max}$ is presented by:

$$V_{max} = \max_k(\max_{ij}(V_{ij}^k)), k \in [1..H] i \in [1..N], j \in [1..L] \tag{16}$$

where $H$ is the number of the used 3D surfaces.

This equation shows that when the variation between two correspondent points tends to reach $V_{max}$ the weight $W_{ij}$ of $p_{ij}$ tends to zero.

Thus, the novel transformation estimation should find a solution to the least squares defined by:

$$\hat{T} = \operatorname*{argmin}_{T} \sum_i \sum_j W_{ij}^2 e_{ij}^2 \tag{17}$$

Seen that $T$ is a rigid transformation, it can be decomposed on rotation and translation. Therefore, it can be defined as follows

$$\hat{T} = \underset{T}{\operatorname{argmin}} \sum_i \sum_j W_{ij}^2(\|p_{ij}^{S_2} - T(p_{ij}^{S_1})\|^2) \tag{18}$$

The translation between the two sets of points is defined by:

$$\hat{t} = C_{S_2} - R C_{S_1} \tag{19}$$

where $C_{S_2}$ and $C_{S_1}$ are respectively the centroid of $\hat{M}^N(S_1)$ and $\hat{M}^N(S_2)$.

Once the rotation $R$ is determined the translation can be derived. Therefore, we need firstly to estimate the rotation $R$. We place each set of points on its centroid landmark: $p_{C_{ij}}^{S_1} = p_{ij}^{S_1} - C_{S_1}$ and $p_{C_{ij}}^{S_2} = p_{ij}^{S_2} - C_{S_1}$. The optimal rotation is rewritten as follows:

$$\hat{R} = \underset{R}{\operatorname{argmin}} \sum_i \sum_j W_{ij}^2 \|p_{C_{ij}}^{S_2} - R(p_{C_{ij}}^{S_1})\|^2 \tag{20}$$

## 5    Experiments and Discussion

Here, we perform experiments based on the novel version of ICP applied to the reparametrized level curves for the identification scenario. For the experimentation, we used a part of the BU-3DFE database [20]. This portion is composed by 700 faces corresponds to the first magnitude level of the six facial expressions and the neutral face of all the subjects of the database(100 subjects). We run

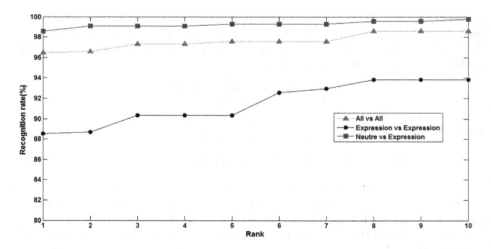

**Fig. 2.** The CMC curve of the proposed approach for the scenarios: All vs All, Expression vs Expression and Neuter vs Expression

the experiments with three protocols: ($i$) The first one is All vs All. It consists on the comparison of each face of the database to all the others. ($ii$) Expression vs Expression is the second protocol. This one corresponds to the comparison between each expression of the database and all the other expressions. ($iii$) Neuter vs Expression protocol is used to compare each 3D neutral face with the 3D faces with expression. Figure 2 shows the Cumulative Matching Curves of the proposed 3D representation under the three protocols cited above. The obtained rank-one recognition rates are about 96.48% for All vs All protocol, 88.53% for Expression vs Expression and 98.65% for Neuter vs Expression.

## 6 Conclusion

We introduced in this work a new approach for the recognition of the 3D faces. This approach consists on a novel robust version of the ICP algorithm. This proposed ICP is based on the three polar representation proposed in [9] and it is adopted to the variation of the facial expression on the 3D faces. The obtained rates for the three protocols of the identification scenario show the performance of the proposed framework.

We propose in the future work to experiment the proposed approach on the standard database of 3D faces FRGCV2. We intend also to compare the proposed ICP with ICP's variants.

## References

1. Paquet E., Rioux M.: A query by content system for three-dimensional model and image databases management. In: The 17th conference on Image and Vision Computing, pp. 157–166 (1999)
2. Osada, R., Funkhouser, T., Chazelle, B., Dobkin, D.: Shape distributions. ACM Trans. Graph. **21**(4), 807–832 (2002)
3. Shinagawa, Y., Kunii, T.-L., Kergosien, Y.-L.: Surface coding based on morse theory. IEEE Comput. Graph. **11**, 66–78 (1991)
4. Ganguly, S., Bhattacharjee, D., Nasipuri, M.: 3D face recognition from range images based on curvature analysis. ICTACT J. Image Video Process. **4**(3), 748 (2014)
5. Samir, C., Srivastava, A., Daoudi, M.: Three dimensional face recognition using shapes of facial curves. IEEE Trans. Pattern Anal. Mach. Intell. **28**(11), 1858–1863 (2006)
6. Srivastava, A., Samir, C., Joshi, S.H., Daoudi, M.: Elastic shape models for face anlysis using curvilinear coordinates. J. Math. Imaging Vision **33**(2), 253–265 (2008)
7. Gadacha, W., Ghorbel, F.: A new 3D surface registration approach depending on a suited resolution: application to 3D faces. In: IEEE Mediterranean and Electrotechnical Conference (MELECON), Hammamet, Tunisia (2012)
8. Ghorbel, F., Jribi, M.: A robust invariant bipolar representation for R3 surfaces: applied to the face description: Springer. Ann. Telecommun. **68**(3–4), 219–230 (2013)

9. Jribi, M., Ghorbel, F.: A stable and invariant three-polar surface representation: application to 3D face description. In: WSCG 2014, the 22nd International Conference in Central Europe on Computer Graphics, Visualization and Computer Vision, Republic (2014)

10. Ghorbel, F.: A unitary formulation for invariant image description: application to image coding **53**(5–6), 242–260 (1998). Special issue Annales des telecommunications

11. Ghorbel, F.: Invariants for shapes and movement. Eleven cases from 1D to 4D and from Euclidean to Projectives (French version), Arts-pi edn., Tunisia (2012)

12. Besl, P.J., Mckay, N.D.: A method for registration of 3-D shapes. IEEE Trans. Pattern Anal. Mach. Intell. **14**(2), 239–256 (1992)

13. Sethian, J.A.: A fast marching level set method for monotonically advancing fronts. Proc. Nat. Acad. Sci. **93**, 1591–1595 (1996)

14. Rihani, A., Jribi, M., Ghorbel, F.: A novel accurate 3D surfaces description using the arc-length reparametrized level curves of the three-polar representation. In: WSCG 2016, the 24th International Conference in Central Europe on Computer Graphics, Visualization and Computer Vision, Republic (2016)

15. Ayache, N.: Computer vision applied to 3D medical imagery: results, trends and future challenges. In: Proceedings of the 6th Symposium on Robotics Research. MIT Press, also Inria Tech. (1993)

16. Faugeras, O., Hebert, M.: The representation, recognition and positioning of 3d shapes from range data. In: Proceedings of the 8th International Conference On Artificial Intelligence, Karlsruhe, BRD, pp. 996–1002, August 1983

17. Rigoutsos, I., Hummel, R.: Robust similarity invariant matching in the presence of noise: a data parallel approach. In: Proceedings of the 8th Israeli Conference on Artificial Intelligence and Computer Vision

18. Gueziec, A., Ayache, N.: Smoothing and matching of 3D-space curves. In: Proceedings of the Second Europeen Conference on Computer Vision Santa Maragherita Ligure, Italy, May 1992

19. Bannour, M.T., Ghorbel, F.: Isotropie de la représentation des surfaces; Application à la description et la visualisation d'objets 3D. In: RFIA 2000, pp. 275–282 (2000)

20. Lijun, Y., Xiaozhou, W., Yi, S., Jun, W., Matthew, J.: A 3D facial expression database for facial behavior research. In: The 7th International Conference on Automatic Face and Gesture Recognition, pp. 211–216 (2006)

21. Szeptycki, P., Ardabilian, M., Chen, L.: A coarse-to-fine curvature analysis-based rotation invariant 3D face landmarking, In: The IEEE 3rd International Conference on Biometrics: Theory, Applications, and Systems, BTAS 2009 (2009)

# 3D Nasal Shape: A New Basis for Soft-Biometrics Recognition

Baiqiang Xia[⊠]

Infolab21, School of Computing and Communications,
Lancaster University, Lancaster LA1 4WA, UK
b.xia@lancaster.ac.uk

**Abstract.** In the past 10 years, Soft-Biometrics recognition using 3D face has become prevailing, with many successful research works developed. In contrast, the usage of facial parts for Soft-Biometrics recognition remains less investigated. In particular, the nasal shape contains rich information for demographic perception. They are usually free from hair/glasses occlusions, and stay robust to facial expressions, which are challenging issues 3D face analysis. In this work, we propose the idea of 3D nasal Soft-Biometrics recognition. To this end, the simple 3D coordinates features are derived from the radial curves representation of the 3D nasal shape. With the 466 earliest scans of FRGCv2 dataset (mainly neutral), we achieved 91% gender (Male/Female) and 94% ethnicity (Asian/Non-asian) classification rates in 10-fold cross-validation. It demonstrates the richness of the nasal shape in presenting the two Soft-Biometrics, and the effectiveness of the proposed recognition scheme. The performances are further confirmed by more rigorous cross-dataset experiments, which also demonstrates the generalization ability of propose approach. When experimenting on the whole FRGCv2 dataset (40% are expressive), comparable recognition performances are achieved, which confirms the general knowledge that the nasal shape stays robust during facial expressions.

**Keywords:** Nasal Soft-Biometrics · 3D · Gender/Ethnicity recognition

## 1 Introduction

Human faces presents rich textural and morphological cues for their peers to recognition their demographic group. Since the 90s of the last century, computer vision researchers have been solving the problem of image-based automatic facial Soft-Biometrics recognition, e.g. gender [7–9] and ethnicity [9,10]. While the conventional works usually rely on the color images, a new trend consisting on the use of the 3D face has emerged in the past 10 years. In addition to the merits of invariant to illumination and facial makeups, the 3D modality has also demonstrated superior Soft-Biometric recognition performance in comparison with 2D texture based methods [11,12]. Remarkably, since the appearance of the Face Recognition Ground Challenge Dataset (FRGCv2) [2], abundant

© Springer International Publishing AG 2017
B. Ben Amor et al. (Eds.): RFMI 2016, CCIS 684, pp. 75–83, 2017.
DOI: 10.1007/978-3-319-60654-5_7

approaches and results have been reported based on that. In [5], *Lahoucine et al.* put geodesic lengths of facial curves in boosting scheme for gender recognition with the 466 earliest scans of the FRGCv2 subjects. In [14], *Toderici et al.* explored wavelets for Gender and Ethnicity recognition on FRGCv2 dataset. In [6], *Xia et al.* deployed the DSF features to capture 3D facial symmetry and averageness for Gender classification on FRGCv2. In [16], *Gilani et al.* used the 3D facial landmarks distances of FRGCv2 for gender classification. In [15], *Wang et al.* performed gender classification with 3D coordinates of FRGCv2 meshes. In [17], *Huang et al.* established a common recognition framework for gender and ethnicity classification on FRGCv2, using the Local Circular Pattern (LCP) features on both depth and color scans of FRGCv2.

Despite the successfulness and prevalence of 3D modality, the underlying common thought in all these works [5,6,14–17] is to use 3D face as basis for Soft-Biometric recognition. However, researches from other domains have also revealed that demographic cues exist in each individual facial parts. For example, researchers in *Sexual Dimorphism* (Male/Female differences) [18,19] have found that male faces usually possess more prominent facial features than female faces. Male faces usually have more protuberant noses, eyebrows, more prominent chins and jaws. The forehead is more backward sloping, and the distance between top-lip and nose-base is longer. In the study of the ethnic differences [20], researchers have found that compared to the North America Whites, Asians usually have broader faces and noses, far apart eyes, and exhibit the greatest difference in the anatomical orbital regions (around the eyes and the eyebrows). In the clinical study reported in [19], *Alphonse et al.* have revealed that Caucasians have significantly lower fetal Fronto-Maxillary Facial Angle (FMFA) measurements than Asians. These findings indicate that significant demographic cues exist in facial feature level, such as the nose, the forehead, the mouth, the chin. However, very limited research attention has been given to specific facial parts, and to test their individual capability in revealing the Soft-Biometric traits. In 3D domain, *Han et al.* have proposed to investigate gender recognition with face divided into four arbitrary regions [23]. In [22], *Hu et al.* have integrated the volume and area information of facial parts and constructed a sparse feature for 3D face gender recognition. None of the above two works have actually explored the natural facial parts alone for Soft-Biometric recognition. To the best of our knowledge, there has been only one work proposed by *Yasmina et al.* [21] in 2D domain, which investigates gender classification with 2D facial parts on the FERET dataset. With SVM Classifier, they achieved around 81.50% gender classification rate with chin, each eye and the mouth parts. With the nose part, they achieved 86.40% classification rate, which suggests the nose region is more discriminant for gender recognition.

From the analysis above, we find that the nose region has remarkable discriminant power over gender and ethnicity. There is a lack of research on 3D facial parts based, especially the 3D nose based Soft-Biometrics recognition. Thus, we propose to explore the idea of 3D nasal Soft-Biometric recognition, in particular for gender and ethnicity recognition. Our research is further motivated by the fact that, when using only the nasal region, we could effectively get rid of the influence of hair/glasses occlusions and the facial expressions, which are challenging issues when working with face scans [3].

## 2  Methodology and Contributions

Our approach consists of a 3D feature extraction step followed by a standard machine learning step for Soft-Biometrics classification. Firstly, given a pre-processed 3D face, a 3D sphere centered at the nosetip is built to remain only the nasal region. In the next, a collection of parameterized radial curves emanating from the nosetip are extracted to represent the 3D nasal region. We then build our features with the simple 3D coordinates on interpolated points of each curve. The features are later fed to machine learning techniques, e.g. the Support Vector Machine (SVM), the Random Forest (RF) and the Linear Discriminant Analysis (LDA), to examine their strength in gender and ethnicity recognition. We note that the proposed approach is fully automated, without the need of human interaction in any step. The main contributions of this work are the following:

- We propose the idea of 3D nasal Soft-Biometrics recognition, in particular the first work in the literature for gender and ethnicity recognition. We demonstrate with experiments that the nasal shape contains rich demographic cues for Soft-biometrics recognition.
- With 466 earliest scans of FRGCv2 dataset (mainly neutral), we achieve 91% gender recognition rate and 94% ethnicity recognition rate in 10-fold cross-validation using simple 3D coordinates features, which demonstrate the effectiveness of 3D nasal shape in revealing our gender and ethnicity groups. When testing on the FU3D dataset, we achieve 90.53% gender classification rate and 100% ethnicity classification rate, which shows the generalization ability of the proposed approach.
- With the whole FRGCv2 dataset which involves significant expression changes, we achieve again 91% gender recognition rate and 94% ethnicity classification rate. The results demonstrate that the nasal shape based approach is invariant to expressions. It echoes the idea that the nasal shape is robust to facial expression changes.

The rest of the paper is organized as follows. In Sect. 3, the 3D nasal dataset construction and feature extraction methods are detailed, as well as a preliminary analysis of the features' competence in revealing the two demographic traits. Nasal Soft-Biometrics recognition experiments and analysis are presented in Sect. 4. Section 5 concludes the work and states future directions.

## 3  Feature Extraction and Preliminary Analysis

### 3.1  3D Nasal Dataset Construction and Feature Extraction

As there is no specific 3D nose dataset, we are motivated to define and extract the nasal region from 3D face scans, and construct the 3D nasal dataset by ourselves.

We base our study on the FRGCv2 dataset [2], which has been extensively used in 3D face analysis. This dataset contains 4, 007 near-frontal 3D face scans of 466 subjects. There are 1, 848 scans of 203 female subjects, and 2, 159 scans of 265 male subjects. For ethnicity, 1213 scans of 112 subjects are related to Asian group, and the rest 2, 794 scans of 354 subjects are Non-asian. About 60% of the scans have a neutral expression, and the others show expressions of disgust, happiness, sadness and surprise.

Given a near-frontal 3D face scan in FRGCv2, after preprocessing [3], the face is in normalized position and we are able to detect the nosetip simply according to the closest point to the camera. The nose region is then cropped out by a 3D sphere centered at the nosetip. In practice, we set the sphere radius to be 45 mm, in accordance to the statistics of adult nasal bridge length [1]. Following this, we employ the radial curves extraction technique to represent the 3D nasal mesh. This technique has been widely and successfully used in 3D face related studies [3–5, 13]. In our work, we extract densely 100 radius curves emanating from the nosetip on the nasal mesh by equally angular profile extraction, and represent each radial curve with a resolution of 50 interpolated 3D points. The extracted nasal region contains mainly the nose and the region between upper lip and nose. The most flexible facial parts during expression, such as the mouth and cheek regions, have been effectively excluded. Also, the forehead and eye regions, which would suffer significantly in the presence of hair or glasses occlusions, are also successfully removed. The dense radial curves representation has resulted into a close representation to the original nasal shape, and a simple 3D coordinates based feature extraction, which forms the basis of further investigations in this work.

## 4    Experiments

We adopt the experimental settings proposed in 3D face related study [6], which designs two separate experimental sessions on the FRGCv2 dataset. The first session executes 10-fold cross-validation with the 466 earliest scans of FRGCv2 subjects. This subset contains mainly neutral scans. We term this as ***Expression-Dependent*** setting. The second session performs 10-fold cross-validation on the whole FRGCv2, for which 40% are expressive. This setting allows to test the robustness of the proposed approach with facial expressions challenges. In our study, this setting also allows to examine the general belief that nasal region is robust to facial expressions. We term this as ***Expression-Independent*** setting. In both settings, no subject is used in both training and testing stages of the same experiment. Experimental results from 3 classifiers, namely the Support Vector Machine (SVM) with linear kernel, the Random Forest (RF), and the Linear Discriminant Analysis (LDA), are reported in the following.

### 4.1   Expression-Dependent Gender and Ethnicity Recognition

Expression-Dependent gender classification results are depicted in the left of Fig. 1. The height of each bar signifies the mean classification rate, and the red line shows the standard deviation in the 10 folds. All the classifiers achieve >85% mean classification rate. With the linear-kernel SVM classifier, we achieve 90.99% gender classification rate, with a standard deviation of 4.29%. This result is further detailed as confusion matrix in Table 3. The recognition rates for male (91.63%) and female (90.15%) are very close, which means the proposed app-roach performs in balance way between two gender groups. The above results demonstrate that the nasal shape contains rich demographic cues for gender recognition, and our proposed approach can perform effective nasal shape based gender recognition. To confirm these findings, we also performs **cross-dataset** experiment: training with 466 scans of FRGCv2 and testing on the Florence University 3D face dataset (FU3D [24]). The FU3D dataset contains 53 neutral-frontal face scans of 53 Caucasian (Non-asian) subjects. For gender, 14 of the subjects are female and 39 are male. The results are shown in the right of Fig. 1. Despite the significant difference in data acquisition techniques (A laser rangefinder is used in FRGCv2 and a multi-cameras stereo system is used for FU3D), the linear-kernel SVM classifier achieves 90.57% gender classification rate in the cross-dataset experiment. It also demonstrates the generalization ability of our proposed approach in nasal shape based gender classification.

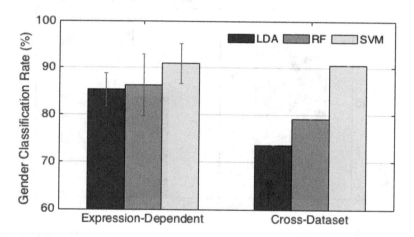

**Fig. 1.** Expression-Dependent gender classification. (Expression-Dependent: experi-ment on 466 FRGCv2 scans; Cross-Dataset:train on 466 FRGCv2 scans, test on FU3D)

For Expression-Dependent ethnicity recognition, the results are depicted in the left of Fig. 2. All the classifiers achieve >90% mean classification rate. With the linear-kernel SVM classifier, we achieve 94.21% ethnicity classification rate, with a standard deviation of 4.06%. Details of this result is presented in Table 4. The recognition rates for Asian (88.39%) and Non-asian (96.05%) are effective, but considerably different. We assume that the unequal number of available scans for Asians (112 scans) and Non-asians (354 scans) accounts largely for this imbalanced performance. In conclusion, the above results highlight that

**Table 1.** Confusion matrix of Expression-Dependent gender classification (T: Truth, P: Prediction)

| T/P | Male | Female | ♯ Scans |
|---|---|---|---|
| Male | 91.63% | 8.47% | 203 |
| Female | 9.85% | 90.15% | 263 |

**Table 2.** Confusion matrix of Expression-Dependent ethnicity classification (T: Truth, P: Prediction)

| T/P | Asian | Non-asian | ♯ Scans |
|---|---|---|---|
| Asian | 88.39% | 11.61% | 112 |
| Non-asian | 3.95% | 96.05% | 354 |

the nasal shape contains rich cues for ethnicity recognition, and the proposed ethnicity recognition approach is effective. Similar to gender, we perform cross-dataset ethnicity recognition with the FU3D dataset. As shown in the right of Fig. 2, the Random Forest classifier gets 100% of the 53 scans correctly classified, and the linear-kernel SVM results in 98.11% ethnicity classification rate. This result again demonstrates the generalization ability of our approach in nasal shape based ethnicity classification.

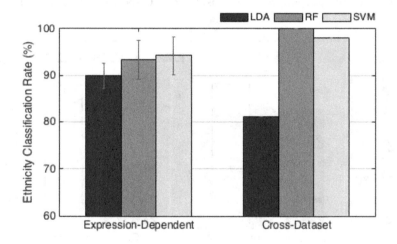

**Fig. 2.** Expression-Dependent Ethnicity Classification. (Expression-Dependent: experiment on 466 FRGCv2 scans; Cross-Dataset:train on 466 FRGCv2 scans, test on FU3D)

### 4.2 Expression-Independent Gender and Ethnicity Recognition

In this section, experiments are carried out on the whole FRGCv2 dataset, for which about 40% scans are expressive. As shown in Fig. 3 (A), for the 3 classifiers, the gender classification rates stay comparable to the Expression-Dependent setting, but the standard deviations decrease significantly. The linear-kernel SVM classifier achieves 91.26% gender classification rate, with a standard deviation of 1.59%. The confusion matrix in Table 3 shows balanced recognition performance for both male (91.71%) and female (91.07%). For ethnicity recognition, results

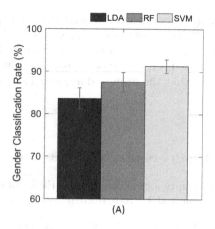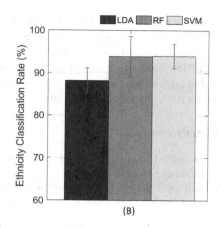

**Fig. 3.** Gender and Ethnicity recognition results under Expression-Independent setting. (A) Results for gender classification; (B) Results for ethnicity classification

**Table 3.** Confusion Matrix of Expression-Independent Gender Classification (T:Truth, P:Prediction)

| T/P | Male | Female | ♯ Scans |
|---|---|---|---|
| Male | 91.71% | 8.29% | 1847 |
| Female | 8.93% | 91.07% | 2160 |

**Table 4.** Confusion matrix of Expression-Independent Ethnicity Classification (T:Truth, P:Prediction)

| T/P | Asian | Non-asian | ♯ Scans |
|---|---|---|---|
| Asian | 88.43% | 11.57% | 1213 |
| Non-asian | 3.66% | 96.34% | 2794 |

are depicted in Fig. 3 (B). The linear-kernel SVM achieves 93.96% Asian and Non-asian classification rate. The confusion matrix in Table 4 shows the performance for Asian (88.43%) is lower than the performance of Non-asian (96.34%). However, these results are no worse than those from the Expression-Dependent experiments. In conclusion, the above results demonstrate that our proposed approaches for recognizing these two Soft-Biometric traits are robust to facial expression challenges.

## 5 Conclusion

In this work, we have proposed the idea of 3D nasal Soft-Biometrics, in particular for gender and ethnicity classification. The nasal shape is free from hair and glass occlusions, and also stays relatively rigid and robust with expression changes. Using the simple 3D coordinates features, under the Expression-Dependent setting, the cross-validation experiments carried out on FRGCv2 dataset have achieved 91% gender recognition rate and 94% ethnicity recognition rate, which confirm the usefulness of nasal shape, and the effectiveness of proposed approach. Cross dataset experiments tested on the FU3D dataset have reached 90.5% gender recognition rate and 100% ethnicity recognition rate, which demonstrate the generalization ability of proposed approach on other dataset. When testing with the Expression-Independent setting on the Whole FRGCv2 dataset, for both gender and ethnicity, comparable performances have been achieved in terms of recognition rate, and higher performances have been observed considering the standard deviation. These results demonstrate that the proposed approach stays stable with expression challenges.

## References

1. 3D Facial Norms Summary Statistics. http://www.facebase.org/facial_norms/summary/#nasalbdglength
2. Phillips, P.J., et al.: Overview of the face recognition grand challenge. IEEE Computer Society Conference on Computer Vision and Pattern Recognition (CVPR 2005), vol. 1. IEEE (2005)
3. Drira, H., Amor, B.B., Srivastava, A., Daoudi, M., Slama, R.: 3D face recognition under expressions, occlusions, and pose variations. IEEE Trans. Pattern Anal. Mach. Intell. **35**(9), 2270–83 (2013)
4. Amor, B.B., Drira, H., Berretti, S., Daoudi, M., Srivastava, A.: 4-D facial expression recognition by learning geometric deformations. IEEE Trans. Cybern. **44**(12), 2443–2457 (2014)
5. Ballihi, L., Amor, B.B., Daoudi, M., Srivastava, A., Aboutajdine, D.: Boosting 3-D-geometric features for efficient face recognition and gender classification. IEEE Trans. Inf. Forensics Secur. **7**(6), 1766–1779 (2012)
6. Xia, B., Amor, B.B., Drira, H., Daoudi, M., Ballihi, L.: Combining face averageness and symmetry for 3D-based gender classification. Pattern Recogn. **48**(3), 746–758 (2015)
7. Ng, C.B., Tay, Y.H., Goi, B.M.: Vision-based human gender recognition: A survey. arXiv preprint arXiv:1204.1611 (2012)
8. Guo, G.: Human age estimation and sex classification. In: Shan, C., Porikli, F., Xiang, T., Gong, S. (eds.) Video Analytics for Business Intelligence. SCI, vol. 409, pp. 101–131. Springer, Heidelberg (2012). doi:10.1007/978-3-642-28598-1_4
9. Han, H., Otto, C., Liu, X., Jain, A.: Demographic estimation from face images: human vs machine performance. IEEE Trans. Pattern Anal. Mach. Intell. **37**(6), 1148–1161 (2014). http://ieeexplore.ieee.org/xpl/downloadCitations
10. Fu, S., He, H., Hou, Z.G.: Learning race from face: a survey. IEEE Trans. Pattern Anal. Mach. Intell. **36**(12), 2483–2509 (2014)
11. Xia, B., Amor, B.B., Huang, D., Daoudi, M., Wang, Y., Drira, H.: Enhancing gender classification by combining 3D and 2D face modalities. In: 21st European Signal Processing Conference, pp. 1–5 (2013)
12. Zhang, W., Smith, M., Smith, L., Farooq, A.: Gender recognition from facial images: 2D or 3D? J. Optical Soc. Am. A **33**(3), 333–344 (2016). ISSN 1084-7529
13. Ezghari, S., Belghini, N., Zahi, A., Zarghili, A.: A gender classification approach based on 3D depth-radial curves and fuzzy similarity based classification. In: Intelligent Systems and Computer Vision (ISCV), pp. 1–6 (2015)
14. Toderici, G., O'malley, S.M., Passalis, G., Theoharis, T., Kakadiaris, I.A.: Ethnicity-and gender-based subject retrieval using 3-D face-recognition techniques. Int. J. Comput. Vision **89**(2–3), 382–391 (2010)
15. Wang, X., Kambhamettu, C.: Gender classification of depth images based on shape and texture analysis. In: Global Conference on Signal and Information Processing (GlobalSIP), pp. 1077–1080. IEEE (2013)

16. Gilani, S.Z., Shafait, F., Mian, A.: Biologically significant facial landmarks: how significant are they for gender classification? In: 2013 International Conference on Digital Image Computing: Techniques and Applications (DICTA), pp. 1–8 (2013)
17. Huang, D., Ding, H., Wang, C., Wang, Y., Zhang, G., Chen, L.: Local circular patterns for multi-modal facial gender and ethnicity classification. Image Vis. Comput. **32**(12), 1181–1193 (2014)
18. Bruce, V., Burton, A.M., Hanna, E., Healey, P., Mason, O., Coombes, A., Fright, R., Linney, A.: Sex discrimination: how do we tell the difference between male and female faces? J. Percept. **22**(2), 131–152 (1993)
19. Jennifer, A., Jennifer, C., Jill, C., Philip, S., Andrew, M.: The Effect of Ethnicity on 2D and 3D Frontomaxillary Facial Angle Measurement in the First Trimester, Obstetrics and Gynecology International (2013)
20. Farkas, L.G., Katic, M.J., Forrest, C.R.: International anthropometric study of facial morphology in various ethnic groups/races. J. Craniofac. Surg. **16**(4), 615–646 (2005)
21. Andreu, Y., Mollineda, R.A.: The role of face parts in gender recognition. In: Campilho, A., Kamel, M. (eds.) ICIAR 2008. LNCS, vol. 5112, pp. 945–954. Springer, Heidelberg (2008). doi:10.1007/978-3-540-69812-8_94
22. Hu, Y., Yan, J., Shi, P.: A fusion-based method for 3D facial gender classification. In: 2010 The 2nd International Conference on Computer and Automation Engineering (ICCAE), vol. 5. IEEE (2010)
23. Han, X., Ugail, H., Palmer, I.: Gender classification based on 3D face geometry features using SVM. In: CyberWorlds, pp. 114–118 (2009)
24. Bagdanov, A.D., Del Bimbo, A., Masi, I.: The florence 2D/3D hybrid face dataset. In: Proceedings of the 2011 Joint ACM Workshop on Human Gesture and Behavior Understanding, pp. 79–80 (2011). ISBN 978-1-4503-0998-1

# Towards a Methodology for Retrieving Suspects Using 3D Facial Descriptors

Naoufel Werghi[1(✉)] and Hassen Drira[2]

[1] Khalifa University, Abu Dhabi, UAE
naoufel.werghi@kustar.ac.ae
[2] Institut Mines-Tlcom/Tlcom Lille, Centre de Recherche en Informatique,
Signal et Automatique de Lille (UMR CNRS 9189), Villeneuve-d'Ascq, France

**Abstract.** We propose a first investigation towards a methodology for exploiting 3D descriptors in suspect retrieval in the context of crime investigation. In this field, the standard method is to construct a facial composite, based on witness description, by an artist of via software, then search a match for it in legal databases. An alternative or complementary scheme would be to define a system of 3D facial attributes that can fit human verbal face description and use them to annotate face databases. Such framework allows a more efficient search of legal face database and more effective suspect shortlisting. In this paper, we describe some first steps towards that goal, whereby we define some novel 3D face attributes, we analyze their capacity for face categorization though a hieratical clustering analysis. Then we present some experiments, using a cohort of 107 subjects, assessing the extent to which some faces partition based on some of these attributes meets its human-based counterpart. Both the clustering analysis and the experiments results reveal encouraging indicators for this novel proposed scheme.

**Keywords:** 3D face · Clustering · Face recognition

## 1 Introduction

In criminology and police investigation, facial sketches (called also facial composite) are commonly used in searching and identifying suspects in crimes, in the absence of the suspect(s) photos [14]. In addition to identification, facial composite can be used as additional evidence, to assist investigation at checking leads and to defuse warning of vulnerable population against serial offenders. The sketch of the face used in criminal investigations can be divided into two categories: (a) Legal sketches: these sketches are drawn by forensic artists referring to the description provided by a witness. Judicial Sketches have been used in criminal investigations since the 19th century [16]; (b) Composite sketches: the sketches of faces are rather built using software allowing an operator to select and combine different elements of the face. Composite sketches are increasingly used. It is now estimated that 80% of law enforcement agencies use software to create facial sketches of suspects [16].

© Springer International Publishing AG 2017
B. Ben Amor et al. (Eds.): RFMI 2016, CCIS 684, pp. 84–94, 2017.
DOI: 10.1007/978-3-319-60654-5_8

The current procedure of suspect identification based on witness description as currently adopted by authorities does not yet seem to profit from all the available resources. In particular the face database maintained by legal authorities and which are continuously fed form network of cameras deployed at access control points and public places. Performance-wise, the current procedures suffer from several shortcomings. Legal sketches production is subjective and depends on the artist skills. Facial composite software, while offer comprehensive construction functionalities, they often produce a mismatched outcomes. Moreover, both categories use 2D face reconstruction, which does not accurately reflect the actual 3D shape features of the subject. Recently some methods proposed to match the face sketch to mugshots (photos of person taken after being arrested) [15,20] and composite sketches to mugshots [12,22]. In both of these two schemes, witness description goes through a human interpretation stage, namely the expert artist for face sketch, and the software operator for the composite sketch. Both face sketch and composite sketch are therefore subjected to reconstruction error. Time required to generate the sketch can be problematic for cases requiring immediate investigation.

More recently, a face retrieval approach trend was pioneered by Klare et al. [8]. In this approach a set of textual description of the suspect face are used to interrogate a face database and retrieve a set of potential suspect(s).

The primary investigation, conducted with 2D images, showed that such scheme achieve retrieval performance comparable to the sketch-based part, and has the potential of improving further the accuracy through fusion.

In the this work we proposes investigating a 3D facial image approach of this scheme and capitalizing on the intrinsic advantages the 3D facial images especially with the regard of the shape information. Indeed, a large number of pertinent facial attributes emanate from the facial shape, that one can notice when contemplating a face. These include global attributes (e.g. overall face shap) and local attributes (nose and eye shapes). This approach has higher potential for retrieving facial trait and features that are not preserved in 2D images because of the loss of geometry by projection. The practical deployment of this approach in surveillance and investigations scenarios involving large date sets require an automatic annotation of these last. In this scope, the paper proposes first steps towards this objective.

## 2    Shape-Based 3D Face Description

### 2.1    Face Kernel

The face kernel is a concept inspired from the "starshapeness" framework [10] in which the kernel (Kern) of a surface is the space (e.g. the set of points) from which the interior of whole surface is visible. It was firstly proposed by Werghi [17] for the purpose of spherical mapping and alignment of facial surfaces. Here we suggest the face kernel a global facial descriptor as mean for describing global properties of the facial surface. This intuition behind this suggestion is that the face kernel reflects the convexity of a surface. For instance the kernel of convex surface, (plane or sphere) is the whole space encompassed by that surface. Its counterpart for a non-convex surface will be much more reduced depending on the amount of self-occlusion inferred by protrusions and cavities in this surface. Figure 1 depicts some surface kernels illustrating such difference.

**Fig. 1.** Examples of kernel for a planar patch, ellipsoid patch and prismatic convex surface. kernels are cut from the left side because of limited space.

We suggest that he size of the face kernel has the potential of reflecting the facial landscape features in terms of the extent of protrusion and concavities that it does exhibit.

**Face Kernel Construction.** For a traingual mesh manifold surface $\mathcal{S}(V, F)$, where $V$ and $F$ refer to the vertices and the facets, respectively, we can demonstrate that the kernel of the surface $\mathcal{S}$ as follows [17]

$$Kern(\mathcal{S}) = \bigcap_{i=1}^{n} \mathcal{H}_i \tag{1}$$

Where $n$ is the number of facets in the mesh surface $\mathcal{S}$, and $\mathcal{H}_i$ is the negative half spaces associated to the plane containing the triangular facet $f_i$. For an oriented plane, the negative half space if the set of points that fall beneath that plane, as opposite to positive half space that include points which are above that plane. The above definition allows an iterative construction of the surface kernel in a space-carving fashion by initializing it to the whole space then successively discarding from it the positive half space associated to the facet $f_i$, as presented in the following algorithm

**Kernel construction**

>   *Generate a 3-D Grid of points $\mathcal{G}$ encompassing the surface $\mathcal{S}_n$.*
>
>   $Ker(\mathcal{S}_n) \longleftarrow \mathcal{G}$
>
>   **For** *each facet $t_i$*
>
>>      *Find, in $Ker(\mathcal{S}_n)$, the set of points $Y_i$ that lie in the half-space $\mathcal{H}_i$*
>>
>>      $Ker(\mathcal{S}_n) \longleftarrow Y_i$
>
>   **End For**

**End Kernel construction**

Figure 2(a–c) depicts different stages of the kernel construction of a facial surface.

Practically, there is a need to check the integrity of the normal across all the facial mesh surface (to avoid wrongly flipped facet normals) and to apply and optimal smoothing and mesh-regularization of the facial surface to avoid the kernel being affected by mesh artifacts as we will see in the experiments.

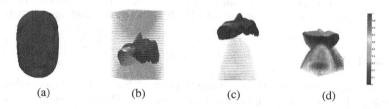

|          |          |          |          |
|:--------:|:--------:|:--------:|:--------:|
| (a)      | (b)      | (c)      | (d)      |

**Fig. 2.** Kernel construction: (a) Spherical cropping of the facial surface, (b) initial kernel composed a dense set of points encompassing the facial surface. (c) the final kernel. (d) The "Goodness of Visibility" computed at teach point in the kernel.

**The Goodness of Visibility.** A complementary aspect to the kernel concept is what we call the "goodness of visibility" of the surface, which we define according to the rule of thumb: A surface is best viewed when the line of sight reaches it perpendicularly. While the interiro of a surface is visible from any point in the kernel, some points allow a better view then others. For example, for sphere surface, for which the kernel is its whole interior, the center is the point having the best view, as any ray fired from this point towards the surface, is colinear with the normal at the point of intersection. For a given point in the kernel, We define the "goodness of visibility" by

$$\mathcal{V} = \frac{1}{K} \int_{\mathcal{S}} \varrho_s ds \qquad (2)$$

where $\varrho_d s$ is scalar product of the unit vector defining the orientation of the ray fired from the kernel point towards the facial surface and the local normal at the interception point. $K$ is a normalizing factor. Figure 2d shows the Goodness of visibility colormapped at each point of the a face kernel. Notice that points that are kernel borders, particularly the closed to the surface, have a less visibility. In contrast with those located in the central zone and at a larger setback distance from the facial surface. These observations fit with the human intuition that a best view of given surface is the one which is centrality and symmetry wihe respect to the surface. Also while it was not an intention of this research we believe, that the concept of the surface kernel (1) and the goodness of visibility derived from it (2) is a novel and an original criterion, expected to be strong competitor to other standard best viewpoint criteria proposed in the literature [4] (Fig. 3).

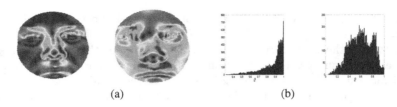

(a)                              (b)

**Fig. 3.** (a): Examples showing mappings of $\varrho$ on the cropped facial surface obtained for the best (maximum) and worst (minimum) $\mathcal{V}$. (b): The corresponding $\varrho$ distributions.

## 2.2   Nasal Profile

In this section we investigate three nasal profiles for human recognition. The first curve is the geodesic path between eyes corners. The geodesic path between nose corners represents the second curve and the third curve (the vertical profile curve) is the geodesic between the mid-eye point and the point lying in the middle of mouth corners. Examples of extracted curves are illustrated in Fig. 4.

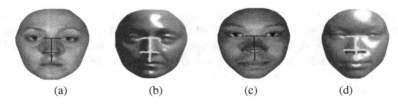

(a)              (b)              (c)              (d)

**Fig. 4.** Illustration of nasal curves: the curves represent the geodesic paths between several facial landmarks.

**Background on Shape Analysis of Profile Curves.** Let $\beta : I \to \mathbb{R}^2$, represent a parameterized curve representing a nasal profile, where $I = [0, 1]$. To analyze the shape of $\beta$, we shall represent it mathematically using the *square-root velocity function* (SRVF) [19], denoted by $q(t)$, according to: $q(t) = \frac{\dot{\beta}(t)}{\sqrt{\|\dot{\beta}(t)\|}}$; $q(t)$ is a special function of $\beta$ that simplifies computations under elastic metric.

Actually, under $\mathbb{L}^2$-metric, the re-parametrization group acts by isometries on the manifold of $q$ functions, which is not the case for the original curve $\beta$. Let's define the preshape space of such curves: $\mathcal{C} = \{q : I \to \mathbb{R}^2 \|\|q\| = 1\} \subset \mathbb{L}^2(I, \mathbb{R}^2)$, where $\| \cdot \|$ implies the $\mathbb{L}^2$ norm. With the $\mathbb{L}^2$ metric on its tangent spaces, $\mathcal{C}$ becomes a Riemannian manifold. Also, since the elements of $\mathcal{C}$ have a unit $\mathbb{L}^2$ norm, $\mathcal{C}$ is a hypersphere in the Hilbert space $\mathbb{L}^2(I, \mathbb{R}^2)$. The geodesic path between any two points $q_1, q_2 \in \mathcal{C}$ is given by the great circle, $\psi : [0, 1] \to \mathcal{C}$, where

$$\psi(\tau) = \frac{1}{\sin(\theta)} \left( \sin((1 - \tau)\theta)q_1 + \sin(\theta\tau)q_2 \right), \tag{3}$$

and the geodesic length is $\theta = d_c(q_1, q_2) = cos^{-1}(\langle q_1, q_2 \rangle)$.

In order to study *shapes* of curves, one identifies all rotations and re-parameterizations of a curve as an equivalence class. Define the equivalent class of $q$ as:

$$[q] = \text{closure}\{\sqrt{\dot{\gamma}(t)}O.q(\gamma(t)), \quad \gamma \in \Gamma\}, \tag{4}$$

where $O \in SO(3)$ is a rotation matrix in $\mathbb{R}^3$.

The set of such equivalence classes, denoted by $\mathcal{S} \doteq \{[q]|q \in \mathcal{C}\}$ is called the *shape space* of open curves in $\mathbb{R}^2$. As described in [19], $\mathcal{S}$ inherits a Riemannian metric from the larger space $\mathcal{C}$ due to the quotient structure. To obtain geodesics and geodesic distances between elements of $\mathcal{S}$, one needs to solve the optimization problem:

$$(O^*, \gamma^*) = argmin_{\gamma \in \Gamma, O \in SO(3)} d_c(q_1, \sqrt{\dot{\gamma}}O.(q_2 \circ \gamma)). \tag{5}$$

Let $q_2^*(t) = \sqrt{\gamma^*(t)}O^*.q_2(\gamma^*(t))$ be the optimal element of $[q_2]$, associated with the optimal re-parameterization $\gamma^*$ of the second curve and the optimal rotation $O^*$, then the geodesic distance between $[q_1]$ and $[q_2]$ in $\mathcal{S}$ is $d_s([q_1], [q_2]) \doteq d_c(q_1, q_2^*)$ and the geodesic is given by Eq. 3, with $q_2$ replaced by $q_2^*$. This representation was previously investigated for biometric [1,5–7] and soft-biometric applications [2,21] based on the face shape.

## 3    Experiments

We conducted a series of experiments aiming at (1) Analyzing the distribution of the kernel size and the Goodness of Visibility to investigate the presence of potential semantic partition; And (2) Searching for some evidence that can support the concordance of these face descriptors with the human perception when categorizing face based on some morphological traits. In the experiments we used a dataset of 105 scans, from the Bosphorus database [18] corresponding to the set of subjects scanned in neutral pose including male and female instance. This data was first-reprocessed to uniform the mesh, and to remove artifacts using Laplacian smoothing.

### 3.1    Clustering Analysis

In the first experiments we conducted a series of Hierarchical clustering on the proposed facial attributes. Different Hierarchical clustering methods Can be investigated [13]. Most of the works of Hierarchical clustering of facial images were related to subject recognition [3,9]. Recently Grant and Flynn investigate Hierarchical clustering beyond subject identification, as to prove the existence of cluster by gender race, and illumination condition. Little or nothing has been done on whole 3D facial images to the best of our knowledge. The goal of this analysis is to explore the extent to which our attributes can form the basis of semantic partition and a meaningful categorization of the facial shapes, and therefore can be adopted to face annotation. We adopted an agglomerative hierarchical clustering using the standard average methods. Other variants such as the single, complete abd Ward [13] could be used as well.

Figure 5a shows the dendrogram of the nasal profiles based classification. We notice that the dendogram exhebits two main distinctive clusters. The examination of these samples Fig. 5b reveals clear different aspects in nasal profiles.

Figure 6(a) shows the dendrograms of the kernel size. We notice that the dendogram exhebits three distinctive and fairly balanced clusters. On the right are three representative samples from the extrema leaves in the tree. The examination of these samples reveals clear dissimilarities aspect in the face morphology. Indeed we can notice that first group show even shape with moderate variation. In the opposite the second group exhibits ample protrusion (nose) and intrusion (eye sockets) marking salient features of the face. We notice in particular the second sample shows a lateral nose deformation. Such feature reduces further the visibility of the surface.

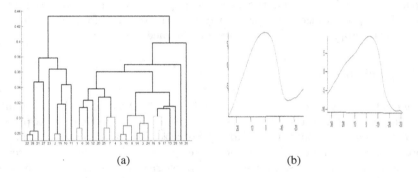

(a)                                    (b)

**Fig. 5.** (a) Nasal profile based dendogram. (b) Representative samples from two extrema clusters of nasal profiles.

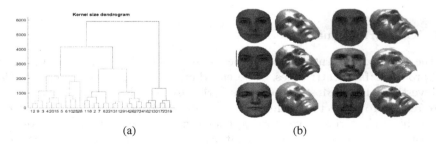

(a)                                    (b)

**Fig. 6.** (a) kernel size dendogram. (b) Representative samples from two extrema clusters

The dendogram of the goodness of visibility is depicted in Fig. 7. Here also we notice three distinctive clusters. As for the kernel size, the three samples of the extrema tree clusters show clear contrast. The first group exhibit rather smooth and even-shaped face, again with an overall flattens aspect. The other group is characterized by a blatant saliency appearance exhibiting significant eye socket intrusion and nose-mouth protrusion with an overall acute shape.

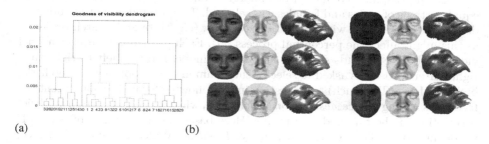

(a)                                           (b)

**Fig. 7.** (a) Goodness of visibility dendrogram. (b) Representative samples from two extrema clusters with their corresponding $\varrho$ mappings.

To have an idea on the range of variation of the kernel size and thew Goodness of visibility we plotted the values of these two descriptors in ascending order (see Fig. 8) for the 105 subjects. From the plots we can notice an range amplitude between the minimum and the maximum values of around 3 and 2 for the kernel size and the Goodness of visibility respectively. We can also notice the clear contrast in the facial shape between the group of the three samples corresponding to the three extrema values in each category.

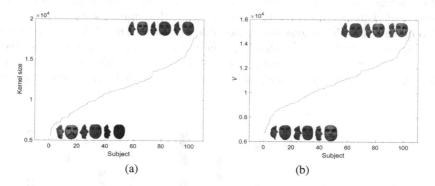

(a)                                           (b)

**Fig. 8.** (a) plots of the ranked kernel size(a) and the Goodness of Visibility (b) for the 100 subjects. Shown also the subjects having the maximum and minimal values for each descriptor.

## 3.2   Human Judgment Matching

This experiment aimed at assessing the extent to which face categorization based on the proposed facial attributes, namely, the kernel size and the goodness of visibility, can match human perception. The experiment was set as follows: A cohort of thirty participants composed of undergrad and postgrad students including equal portions of males and females was selected. The group does not include students that are familiar with the databases faces, (e.g. through research projects) as this might affect the perceptual process [11]. Each participant watches a brief video of about 8 seconds showing the 3D face model rotating left to right then right. Afterwards he is asked to select a choice among three options (a: Too little, b: Somewhat c: Too much) to a question in the following form: To what extent the face looks having a *Face_description* appearance, where *Face_description* is a brief description of the targeted face profile. Here based on the findings of Sect. 3.1, we defined two different profiles, namely: *Profile_1*: Wide, flat, unmarked face; And *Profile_2*: Marked face exhibiting protruding nose, intruding eyes. This procedure is repeated for all the 105 models in the dataset with a pause of 3 min after each judgment.

Scores collected from the participants are averaged for correlation with scores obtained from the kernel size and the goodness of visibility criterion. For that purpose, we mapped score obtained with these two attributes into three sets representing three segments of the their related ranges, and labeled with the three aforementioned options. However, rather than using crisp sets, we considered a mapping to three fuzzy sets as shown in Fig. 9a. This is motivated by the appropriateness of fuzzy rating accommodating comparative judgment and confidence ambiguity characterizing facial description by human [11].

Scores related to the Goodness of Visibility show a slightly better match. The rate of matched scores are reported in Fig. 9(b) and (c) for the kernel size and the goodness of visibility respectively. First we notice that matches with the third set ("Too much" have the largest and reasonable rate (above 80%). Less matches are obtained for the first set ("Too little"), whereas the middle set shows a relatively low matches. Considered relatively to each other, the matching scores give some indication that both face descriptors concord well with human perception for

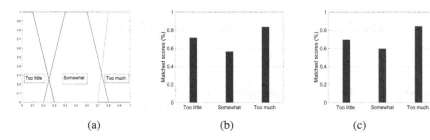

(a)                          (b)                          (c)

**Fig. 9.** (a) Fuzzy sets associated with the three judgment options. (b) matched score rate for the kernel size, (c) matched score for the Goodness of Visibility.

assigning to (and with a less degree rejection from) the aforementioned profiles. This there is some evidence that significant values of these descriptors can be utilized for labeling subject with these profiles.

## 4    Conclusion and Discussion

In this paper, we have presented a novel approach for using 3D facial image for retrieving suspects based on witness description. We proposed two global facial descriptors for categorizing facial morphology. The clustering analysis, we performed, seem providing an encouraging indication about the plausibility of these tools for a semantic subject partition that can be verbally described, and thus having a potential to be utilized for annotation. The experiment assessing the extent to which human perception can meet face categorization based on the proposal global descriptors revealed positive trend in this regard, and confirms further the utility of 3D images,

 While it is true that the dataset we used is not exhaustive and does not encompass the full spectrum of face morphology (little presence of far east ethnecity), the approach we proposed remain, in our opinion, valid for a more diverse set. To accommodate this diversity, there is a need to consider, in addition to other global attributes, local face attributes reflecting pertinent traits, such as nose and mouth shapes and size of the eyes. The next step in our work is to integrate the nasal morphology and work out related descriptors that can manage the wide spectrum of facial nose. This work developed in [6] can provide appropriate guidance.

## References

1. Ben Amor, B., Drira, H., Ballihi, L., Srivastava, A., Daoudi, M., Daoudi, M.: An experimental illustration of 3D facial shape analysis under facial expressions. Annales des Télécommunications **64**(5–6), 369–379 (2009)
2. Ben Amor, B., Drira, H., Berretti, S., Daoudi, M., Srivastava, A., Srivastava, A.: 4-D facial expression recognition by learning geometric deformations. IEEE Trans. Cybern. **44**(12), 2443–2457 (2014)
3. Antonopoulos, P., Nikolaidis, N., Pitas, I.: Hierarchical face clustering using sift image features. In Proceedings of the IEEE Symposium on Computational Intelligence in Image and Signal Processing, pp. 325–329 (2007)
4. Secord, A., Lu, J., Finkelstein, A., Singh, M., Nealen, A.: Perceptual models of viewpoint preference. ACM Trans. Graph. **30**, 1–13 (2011)
5. Drira, H., Ben Amor, B., Daoudi, M.M., Srivastava, A.: Pose and expression-invariant 3D face recognition using elastic radial curves. In: British Machine Vision Conference, pp. 1–11 (2010)
6. Drira, H., Ben Amor, B., Srivastava, A., Daoudi, M.: A riemannian analysis of 3D nose shapes for partial human biometrics. In: International Conference on Computer Vision, pp. 2050–2057 (2009)
7. Drira, H., Ben Amor, B., Srivastava, A., Daoudi, M., Slama, R., Slama, R.: 3d face recognition under expressions, occlusions, and pose variations. IEEE Trans. Pattern Anal. Mach. Intell. **35**(9), 2270–2283 (2013)

8. Klare, B.F., et al.: Suspect identification based on descriptive facial attributes. In: Proceedings of the IEEE/IAPR International Joint Conference on Biometrics, pp. 1–8 (2014)

9. Fan, W., Yeung, D.Y.: Face recognition with image sets using hierarchically extracted exemplars from appearance manifold. In: Proceedings of the IEEE 7th International Conference on Automatic Face and Gesture Recognition, pp. 177–192 (2007)

10. Torzanos, F.A.: The points of local nonconvexity of starshaped objects sets. Pac. J. Math. **11**, 25–35 (1982)

11. Frowd, C.: Craniofacial identification. In: Wilkinson, C., Rynn, C. (eds.) Craniofacial Identification, pp. 42–56 (2012)

12. Han, H., Klare, B., Bonnen, K.: Matching composite sketches to face photos: a component based approach. IEEE Trans. Inform. Forensics Secur. **8**, 191–204 (2013)

13. Jain, A., Dubes, R.C.: Algorithms for Clustering Data. Prentice- Hall Inc., Upper Saddle River (1988)

14. Jain, A., Klare, B., Park, U.: Face matching and retrieval in forensics applications. IEEE Multimedia **19**, 20–28 (2012)

15. Klare, B., Li, Z., Jain, A.: Matching forensic sketches to mug shot photos. IEEE Trans. Pattern Anal. Mach. Intell. **33**, 639–646 (2011)

16. Mcquiston, D., Topp, L., Malpass, R.: Use of facial composite systems in us law enforcement agencies. Psychol. Crime Law **12**, 505–517 (2006)

17. Werghi, N.: The 3D facial kernel: Application to facial surface spherical mapping and alignment. In: Proceedings of the IEEE Conference Systems, Men and Cybernetics, pp. 1777–1784 (2010)

18. Savran, A., Alyüz, N., Dibeklioğlu, H., Çeliktutan, O., Gökberk, B., Sankur, B., Akarun, L.: Bosphorus database for 3D face analysis. In: Proceedings of the First COST 2101 Workshop on Biometrics and Identity Management, May 2008

19. Srivastava, A., Klassen, E., Joshi, S.H., Jermyn, I.H.: Shape analysis of elastic curves in euclidean spaces. IEEE Trans. Pattern Anal. Mach. Intell. **33**(7), 1415–1428 (2011)

20. Wang, X., Tang, X.: Face photo-sketch synthesis and recognition. IEEE Trans. Pattern Anal. Mach. Intell. **31**, 1955–1967 (2009)

21. Xia, B., Ben Amor, B., Drira, H., Daoudi, M., Ballihi, L., Ballihi, L.: Combining face averageness and symmetry for 3d-based gender classification. Pattern Recognit. **48**(3), 746–758 (2015)

22. Yuen, P., Man, C.: Human face image searching system using sketches. IEEE Trans. SMC Part A Syst. Humans **37**, 493–504 (2007)

# Video and Motion Analysis

# Key Frame Selection for Multi-shot Person Re-identification

Mayssa Frikha[1](✉), Omayma Chebbi[2], Emna Fendri[2],
and Mohamed Hammami[2]

[1] MIRACL-FSEG, Sfax University, Road Aeroport Km 4, Sfax, Tunisia
`frikha.mayssa@hotmail.fr`
[2] MIRACL-FS, Sfax University, Road Sokra Km 3, Sfax, Tunisia
`mouka.lfi@gmail.com`, `fendri.msf@gnet.tn`, `mohamed.hammami@fss.rnu.tn`

**Abstract.** Typical person re-identification approaches rely on a single image to model the visual appearance characteristics for each target. The performance of these systems is very limited as they ignore the immense amount of video data produced by the practical surveillance systems. In this paper, we present a novel multi-shot person re-identification approach based on key frame selection. We propose to conduct a global appearance signature by automatically selecting a set of representative appearance images depicting the different body postures from the target's trajectory. Then, these images will be modeled into a global appearance signature to perform the re-identification task based on set matching strategy. The robustness of our approach is validated on the challenging HDA+ dataset in contrast to the limitations of existing approaches.

**Keywords:** Person re-identification · Multi-shot approach · Key frame selection · Global appearance signature · Histogram of Oriented Gradients · Set matching

## 1 Introduction

Person Re-identification (Re-ID) has recently attracted more and more research interest. It presents a fundamental task for the multi-camera automated video surveillance systems. This task aims at identifying a target captured at different times and/or locations, considering a large set of candidates [1]. This problem has widespread applications such as tracking criminals, analyzing suspect movements, and finding missing people, etc.

A typical Re-ID system is divided into two main steps: (1) extracting an appearance signature $AS_{p_i}$ for a given probe person image $p_i$, then (2) matching it by using a similarity metric against a gallery set $G = \{g_1, ..., g_N\}$, and finally the most similar $g_p$ is assigned to the probe image. However, modeling people's appearance is a paramount and challenging problem because people are often monitored at low resolution, under occlusions, with different lighting conditions,

© Springer International Publishing AG 2017
B. Ben Amor et al. (Eds.): RFMI 2016, CCIS 684, pp. 97–110, 2017.
DOI: 10.1007/978-3-319-60654-5_9

viewpoints and poses. However, conventional biometric features such as face or iris are inappropriate due to the uncontrolled acquisition conditions and insufficient image details for extracting robust biometric features [5]. Instead, bodily human features in term of cloths and carried objects present more reliable characteristics for modeling visual aspect of people. State-of-the-art approaches have mainly focused on modeling people's appearance by extracting discriminative and robust visual characteristics. In the literature, appearance-based approaches can be further divided into two groups: (1) single-shot approaches and (2) multi-shot approaches.

Typical person re-identification approaches extract visual characteristics from a single image depicting the target's appearance [13,18–22]. It is worth noting that the literature abounds with single-shot approaches as it represents the simplest and most general case. However, the performance of these systems is limited due to unavailability of tracking information. Obviously, we risk to extract a blurred appearance signature due to the large intra-class variation (e.g. pose and viewpoint variation, partial occlusion).

In the real-world video surveillance systems, the tracking algorithm provides a multiple frames depicting the same person during his movement in front of the camera view. The multi-shot re-identification approaches exploit a multiple images of each target to model an appearance signature [2–4,6–10]. These approaches aim to extract a more complete and invariant signature. However, these algorithms require an additional step that aims to automatically select a set of appearance images which will be considered in the re-id task. Different image selection methods have been proposed. Bird et al. [2] have modeled people by the median color value of each body part accumulated over different frames. In addition, [3,4] have randomly chosen 5 samples for each person. In this case, the selected images risk to be redundant and/or contains uncompleted appearance features (e.g. partially occulted). [6–8] have applied an unsupervised gaussian clustering method [11] to the HSV histograms of the people images. Then, they randomly selected one image from each cluster. The re-id is carried out by considering 2, 5, and 10 images for each target. Bak et al. [9] have proposed to cluster the trajectory based on the body's estimated pose using 3D scene information. Then, they have generated a signature for every different appearance. However, the calibration information of each camera is needed, which is not always available. Wang et al. [10,23] have selected video fragments from image sequences based on a combination of HOG3D features and optic flow energy profile over each image sequence, which is a computational cost for real applications.

However, due to the uncontrolled acquisition conditions for video surveillance systems, several challenges need to be addressed: (1) the frame rate may be different from one camera to another, (2) the length of person's trajectory is not a constant, and (3) the unpredictable people's path may cause different body postures. So, it is irrelevant to extract the representative appearance images uniformly for each person.

In addition, video surveillance systems generate a vast quantity of video sequence that contains an immense amount of information. So, it is important to take advantage as much as possible from these data. In light of these short-comings, we propose a new multi-shot person re-identification approach based on key frame selection. We propose to automatically select a set of representative appearance images that depict the different body posture variations for each target adaptively. The key idea is to eliminate the redundancy and noisy image from the target's trajectory. Then, these selected images will be modeled into a global appearance signature. Next, the matching score is calculated by the average of all pairwise distances as the probe and gallery sets are made of multiple images for each subject instead of a single image as the traditional state-of-the art approaches.

The remainder of the paper was organized as follows. The different steps of the proposed approach were described in Sect. 2. The experimental results and discussions were reported in Sect. 3. Finally, our conclusions were given in Sect. 4.

## 2 Proposed Approach

In this section, we describe the proposed multi-shot person re-identification app-roach in multi-camera videosurveillance system based on key frame selection. Figure 1 shows the framework of the proposed approach. In general, the single camera tracking algorithm produces a collection of image sequence $Q_{p_i,c_j}$ that estimates the trajectory of each person $p_i$, $i \in 1..P$ where $P$ is the number of people, in the camera view $c_j$, $j \in 1..C$ where $C$ is the number of cameras. Each image sequence $Q$ is defined by a set of consecutive images $I$ as $Q = \{I_1, ..., I_S\}$, where $S$ is the number of frames. It must be noted that the value of $S$ is not a constant and depends from the length of person's trajectory.

The first step, namely *Key Frame Selection*, automatically selects a small set of representative appearance images $Q'_{p_i,c_j} = \{I_m, ..., I_n\}$ depicting the different body posture variation during his movement in front of the camera view. The main idea is to discard the redundant and noisy images from the person's tra-jectory, and, to conduct a multi-appearance presentation. We start by modeling the local silhouette contour variations of the person's trajectory by extracting Histogram of Oriented Gradients descriptor. Then, we automatically detect the appearance transitions, and, a representative image for each segment is selected.

The second step, namely *Appearance Signature Extraction*, models the dis-criminative visual characteristics of the selected images $Q'_{p_i,c_j}$ into a Global Appearance Signature $GAS_{p_i,c_j}$ for each person $p_i$. It starts by alleviating the lighting conditions of the uncontrolled acquisition conditions by adopting a pre-processing step based on Grayworld normalization [17]. Then, we extract the Multi-Channel Co-occurrence Matrix [13] descriptor which encodes both color and texture visual features. These feature vectors are gathered into a global appearance signature $GAS_{p_i,c_j} = \{AS_m, ..., AS_n\}$.

Finally, *Appearance Signature Matching* step compares a specified probe per-son $p_i$ against a gallery set $G = \{GAS_1, ..., GAS_N\}$. A set matching strategy

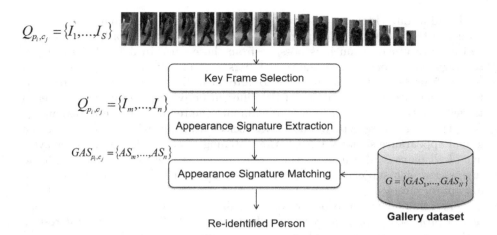

**Fig. 1.** Proposed approach for multi-shot person re-identification based on key frame selection.

based on the average of all pairwise distances is adopted to compute the similarity score between the corresponding probe and the gallery global appearance signatures.

## 2.1  Key Frame Selection

Usually, video surveillance systems generate a vast quantity of video sequence that contains an immense amount of information. However, the temporal correlation between consecutive images may produces redundant information. In addition, cameras are usually mounted above people's head (*e.g.* fixed on the roof), some body parts may be truncated. Hence, these images contain uncompleted appearance information which is irrelevant to be considered in appearance signature extraction step.

Our goal is to benefit from the available data by automatically selecting a set of representative person images, $Q'_{p_i,c_j} = \{I_m, ..., I_n\}$, from the target's trajectory (*i.e.* video sequence), $Q_{p_i,c_j} = \{I_1, ..., I_S\}$, in order to detect the different appearance variations (*e.g.* different body postures). This step is based on two sub-steps as presented in Fig. 2: (1) Key Frame Extraction, and (2) Key Frame Filtering. These two sub-steps aim to eliminate the redundancy and the noisy information from the trajectory, respectively.

**Key Frame Extraction.** Our goal is to eliminate the redundancy from the consecutive images presented in a single trajectory $Q = \{I_1, ..., I_S\}$. First, we start by modeling the body's contour variations during the tracking. Then, we automatically segment the trajectory by detecting the posture transition. Next, we select a representative appearance image for each segment part. More details are provided in the following.

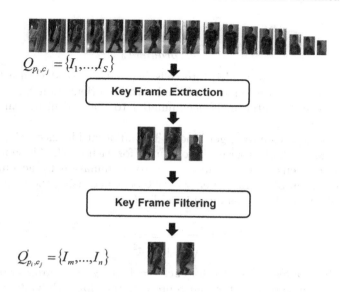

$$Q_{p_i,c_j} = \{I_1,...,I_S\}$$

**Key Frame Extraction**

**Key Frame Filtering**

$$Q'_{p_i,c_j} = \{I_m,...,I_n\}$$

**Fig. 2.** Key frame selection process.

*Modeling body's contour.* The different body postures are modeled by extracting the characteristics of silhouette contours. We opted for Histogram of Oriented Gradient (HOG) features to model each person appearance image $I$. HOG descriptor has been firstly introduced for pedestrian detection problem [12]. It describes the object appearances by analyzing the local distribution of gradients. Extracting the HOG features consists of a series of steps, as shown in Fig. 3.

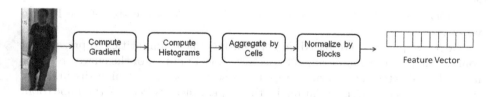

Compute Gradient → Compute Histograms → Aggregate by Cells → Normalize by Blocks → Feature Vector

**Fig. 3.** Overview of the HOG feature extraction process.

First, the gradient of an image is computed by filtering it with two one-dimensional masks $[-1\ 0\ 1]$. For color images, each RGB color channel is processed separately, and pixels assume the gradient vector with the largest norm. After that, the magnitude and the orientation of the gradient are extracted by Eqs. (1) and (2), respectively.

$$|G| = \sqrt{gradient_x^2 + gradient_y^2}\ . \tag{1}$$

$$\theta = arctan\frac{gradient_x}{gradient_y} \, . \tag{2}$$

The gradient image is divided into cells ($C_w \times C_h$ pixels) with a fixed size. For each cell, the orientation is quantized into $C_b$ orientation bins, then, the magnitude in each orientation is accumulated to make a histogram (*i.e.* the feature vector for a cell).

Next, we group cells into larger blocks. The adjacent blocks overlap at half of the block size and local histogram is computed for each block. These histograms are normalized to ensure better invariance to illumination by accumulating a measure of the local histogram energy over blocks and using the results to normalize all cells in the block, as formulated by (3).

$$h = \frac{h}{\sqrt{\|h\|^2 + \varepsilon^2}} \, . \tag{3}$$

This produces a feature vector for each block. These histograms are concatenated into a single feature vector that represents the global HOG description for the whole appearance image.

*Appearance Transition Detection.* We compute the similarity score between the contours of the consecutive frames $I_i$ and $I_{i+1}$. The similarity between HOG features vectors is computed by using Bhattacharyya distance [16], as formulated by (4).

$$Similarity\,(I_i, I_{i+1}) = \sum_{x=1}^{X} \sqrt{HOG^{I_i}\,(x) \times HOG^{I_{i+1}}\,(x)} \, . \tag{4}$$

where:

$HOG^{I_i}$ presents the feature vector for $I_i$, and $X$ is the size of the feature vector.

Next, we plot a curve based on consecutive image similarities to present the contour shape variation during a trajectory. Local minima of the curve are automatically detected, as shown in Fig. 4. These points present high dissimilarity score and hence a high variation in the appearance. Based on their location, we detect the appearance transition and divide the trajectory into several parts $Q_{p_i,c_j} = S_1, ..., S_T$, where $T$ is the number of detected segment which is not a constant. Each part models a posture variation. Clusters with a lower image number ($<= 3$ in our experiments) are merged with the previous cluster.

*Frame Extraction.* For each segment $S_i$, we select a representative appearance image. We take the middle image in each cluster as a key frame.

**Key Frame Filtering.** In general, surveillance cameras are usually mounted above people's head (*e.g.* fixed on the roof). In addition, the entry and the exit of the person in the camera field of view is done progressively. In these cases, some body parts may be partially occulted. If we rely on uncompleted

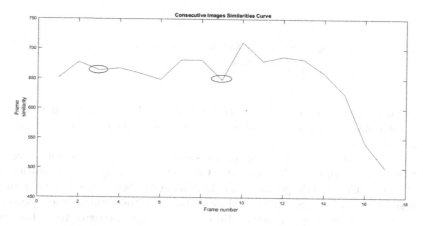

**Fig. 4.** Consecutive images similarities curve.

appearance images to model the visual features, some important characteristics are neglected and omitted. We propose to eliminate the truncated images by analyzing the position of the person image in the camera frame.

Finally, the selected key frame will be modeled in order to extract a discriminative appearance signature to perform the re-id task.

## 2.2 Appearance Signature Extraction

The uncontrolled lighting conditions of video surveillance systems largely affect people's appearance in terms of color correspondence between the different camera views. In this step, we start by alleviating the heterogeneous lighting conditions for each selected image based on Grayworld normalization [17]. This method focuses on measuring clothes colors independently of the light source and camera characteristics, as formulated by Eq. (5). It assumes that the average reflectance in the scene is achromatic.

$$R^{'} = \frac{R}{\mu_R}, G^{'} = \frac{G}{\mu_G}, B^{'} = \frac{B}{\mu_B} \tag{5}$$

where:

$\mu_R$, $\mu_G$, and $\mu_B$ denote, respectively, the average of Red, Green, and Blue image component.

Then, we model the visual characteristics of each corrected image by extracting the Multi-Channel Co-occurrence Matrix (MCCM) [13] descriptor. MCCM descriptor encodes both color and texture visual features. The main advantage it allows to distinguish between people wearing similar colors with different motives. It encodes the information about pixels neighbor by presenting the frequency of a pixel with color value $i$ is adjacent to the pixel with color value $j$ from each body strip $BS_i$, $i \in 1..SBS$, of the size $N \times M$, as expressed in (6).

$$MCCM_{p_i}^{BS}(i,j,k) = \sum_{x=1}^{N}\sum_{y=1}^{M}\begin{cases}1, & BS(x,y,k) = i \wedge BS(x+\triangle_x, y+\triangle_y, k) = j \\ 0, & otherwise\end{cases} \quad (6)$$

where:

$k$ denoted the color component $(H,S,V)$, $\triangle_x$ and $\triangle_y$ are the offset, we calculate $MCCM$ according to the direct neighbor by considering 4 directions ($0°$, $45°$, $90°$, and $135°$).

Then, the four adjacency matrices are summed and converted into a vector. Using the sum of the matrices encodes the distributions of the intensities and totally ignores the relative position of neighboring pixels which is rotation invariant. Also, $HSV$ color space is very close to the way the human brain perceives color and it is known by the separation of the image intensity from the color information which is more robust to lighting variations. In video surveillance systems, the unpredictable camera positions and people's paths aggravates the problems of appearance variation across different viewpoint and pose. The people's appearance in one view is always different from the other views, *i.e.* frontal vs. back views. We have already demonstrated that some body regions are more informative and expose more invariant appearance characteristics against the viewpoint and pose variations [13]. We have automatically selected the salient body stripes from the most adopted body parts model, *i.e.* 6 equalized body stripes, by adopting the Sequential Forward Floating Selection algorithm (SFFS) [14]. As a result, we have proved that regions defined by the upper torso, lower torso and upper legs, $SBS_p = \{UT_p, LT_p, UL_p\}$ are the most stable and informative body parts to capture the salient appearance characteristics in uncontrolled pose and viewpoint conditions. MCCM descriptor has been extracted only from the salient body stripes.

Finally, these feature vectors are gathered into a global appearance signature $GAS_{p_i,c_j} = \{AS_1, ..., AS_n\}$.

### 2.3   Appearance Signature Matching

This step consists in associating each person in the probe set $P = \{p_1, ..., p_N\}$ to the corresponding one in the gallery set $G = \{g_1, ..., g_N\}$. The appearance characteristics of each person is modeled by a global appearance signature for both gallery set, $GAS_P = \{GAS_{p_1}, ..., GAS_{p_N}\}$, and probe set, $GAS_G = \{GAS_{g_1}, ..., GAS_{g_N}\}$. Given a probe person $p_i$, the re-identification is defined as maximum likehood estimation problem [8], as formulated in (7).

$$ID(p_i) = \underset{j \in 1..N}{argmax}\left(d\left(GAS_{p_i}, GAS_{g_j}\right)\right) \quad (7)$$

In our approach, each person is modeled by a global appearance signature that consist of multiple image signatures. To compare two global appearance signatures, we calculate the average of all pairwise distance between two pedestrians as formulated in (8).

$$d\left(GAS_{p_i}, GAS_{g_j}\right) = \frac{1}{M \times N} \sum_{m=1}^{M} \sum_{n=1}^{N} d\left(AS_m^{p_i}, AS_n^{g_j}\right) \tag{8}$$

The distance between two image signatures $AS_m^{p_i}$ and $AS_n^{g_j}$ is computed by using the Bhattacharyya distance [16] from the corresponding MCCM feature vectors, as formulated in (9).

$$d\left(AS_m^{p_i}, AS_m^{g_j}\right) = \frac{1}{\|u\|} \sum_{u \in \{UT, LT, UL\}} d\left(MCCM_u^{p_i}, MCCM_u^{g_j}\right) \tag{9}$$

## 3    Experimental Results

In order to evaluate the performance of our proposed approach, we carried out a series of experiments on the HDA+ dataset [15]. The first experiment validates the choice of HOG parameters for key frame extraction step. The second experiment evaluates the performance of the proposed multi-shot person re-identification approach.

### 3.1    Presentation of HDA+ Dataset

The proposed approach has been experimentally validated on the High-Definition Analytics dataset (HDA+). It has been acquired across 13 indoor cameras in an office distributed over three floors of a research department (e.g. corridors, leisure areas, halls, entries/exits) recording simultaneously for nearly 30 min. Cameras were set to acquire video from different points of view, affecting the geometry of the imaged scene and resulting in different distance ranges. The dataset contains a high degree of viewpoint, pose, and lighting variations. In our experiments, we selected 50 image sequence for 20 people. We have selected those people that reappearing across the different camera views of the network. All images are normalized to $128 \times 64$ pixels.

### 3.2    Performance Validation of the Used HOG Parameters

Our aim is to find the best parameters of the block size and cell size for HOG descriptor that gives the best trajectory segmentation results. The selected people sequences have been manually segmented. Then, we evaluate the segmentation results with the different combination of cell and block size values based on Recall, Precision and F-measure as presented by Eqs. (10), (11), and (12), respectively. Table 1 presents the performance of the different combinations for HOG descriptor.

$$Recall = \frac{number\ of\ correctly\ detected\ transitions}{number\ of\ transitions} \tag{10}$$

$$Precision = \frac{number\ of\ correctly\ detected\ transitions}{number\ of\ detected\ transitions} \tag{11}$$

$$F - measure = 2 \times \frac{Precision \times Recall}{Precision + Recall} \tag{12}$$

**Table 1.** Results of the different cell and block combinations for the HOG descriptor.

| Cell size | 2 × 2 | 2 × 2 | 2 × 2 | 4 × 4 | 4 × 4 | 4 × 4 | 6 × 6 | 6 × 6 | 6 × 6 | 8 × 8 | 8 × 8 | 8 × 8 | 16 × 16 | 16 × 16 | 16 × 16 |
|---|---|---|---|---|---|---|---|---|---|---|---|---|---|---|---|
| Block size | 2 × 2 | 3 × 3 | 4 × 4 | 2 × 2 | 3 × 3 | 4 × 4 | 2 × 2 | 3 × 3 | 4 × 4 | 2 × 2 | 3 × 3 | 4 × 4 | 2 × 2 | 3 × 3 | 4 × 4 |
| Recall | 0.69 | 0.715 | 0.682 | 0.69 | 0.665 | 0.665 | **0.74** | **0.74** | **0.74** | 0.715 | 0.615 | 0.574 | 0.615 | 0.49 | 0.482 |
| Precision | 0.803 | 0.81 | 0.733 | 0.803 | 0.778 | 0.778 | **0.798** | 0.791 | 0.791 | 0.778 | 0.678 | 0.608 | 0.742 | 0.519 | 0.594 |
| F-measure | 0.711 | 0.722 | 0.678 | 0.711 | 0.686 | 0.686 | **0.742** | 0.736 | 0.736 | 0.72 | 0.62 | 0.569 | 0.648 | 0.492 | 0.518 |

As shown in Table 1, the best results are given by Cell size [6 × 6] and Block size [2 × 2]. This combination allows the detection of the fine variations in the silhouette's contour. These parameters will be considered in the rest of the experiments.

### 3.3    Performance Evaluation of Our Proposed Approach

Following the evaluation protocol of the state-of-the-art person re-identification approaches, we randomly select one image sequence to build the gallery set, and one image sequence to build the probe set picked from different camera for each pedestrian. We repeat the experiments for 10 trials and report the average performance. There are a variety of metrics to evaluate the effectiveness of a re-identification system. The results are measured by Cumulative Matching Characteristics curve (CMC) which measures the expectation of the correct match at rank r. We reported the performance at different ranks. Rank-1 accuracy refers the percentage of probe images which are perfectly matched to their corresponding gallery image. In addition, we reported the normalized Area Under the Curve (nAUC) which is derived from the CMC curve. It presents the area under the entire CMC curve normalized over the total area of the graph (a perfect nAUC is 1.0). As we narrow a real-world application, the first ranks in the CMC curve are the most important metrics since they are indicative of how well fully automatic reidentification performs.

In this section, we start by evaluating the adopted method for the set matching for multi-shot re-identification based on key frame selection. It consists of calculating the average of all pairwise distances (*i.e.* Our method). For this evaluation, we compared our solution with the adopted method by Cheng et al. [3] and Farenzena et al. [7] that calculate the maximum of all the sample similarities between two targets (*i.e.* Method A). The re-identification results are given by Table 2 and Fig. 5.

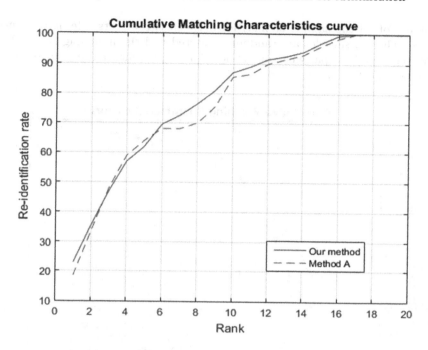

**Fig. 5.** CMC curve for set matching evaluation.

**Table 2.** Re-identification performance for set matching evaluation.

|  | Rank-1 | Rank-2 | Rank-3 | Rank-10 | nAUC |
|---|---|---|---|---|---|
| Our method | **23** | **35.5** | 47 | **87** | **79.63** |
| Method A | 18.5 | 33.5 | **48** | 85.5 | 78.22 |

We can see that our method consistently outperforms method A: re-identification rate at rank 1 for our method is 23% while that of method A is 18.5%. However, the method A depends only on a single sample from each set which can be sensitive to the outliers. As an example, if we compare two pedestrians $p_1$ and $p_2$. $p_1$ wears an open dark coat with a multi-color plaid shirt and a dark trouser. $p_2$ wears a dark shirt and a trouser. If we compare them from the rear view, they will be very similar. But, if we compare them from the frontal view, we can clearly detect the appearance differences. In contrast, our method takes advantage from the different pairwise distances and combine them together.

Then, the second part is dedicated to evaluate the proposed key frame selection method for multi-shot person re-identification (*i.e.* Our method). For this evaluation, we compared our solution with the proposed method by Cheng et al. [3] and Zeng et al. [4]. They have randomly selected 5 images for each person. The matching policy for Cheng et al. [3] (*i.e.* Method B) is based on finding the maximum of all the sample similarities between two targets. Method C refers the case where 5 samples are randomly selected and the matching is based on the average of all the pairwise similarities. The re-identification results are given by Table 3 and Fig. 6.

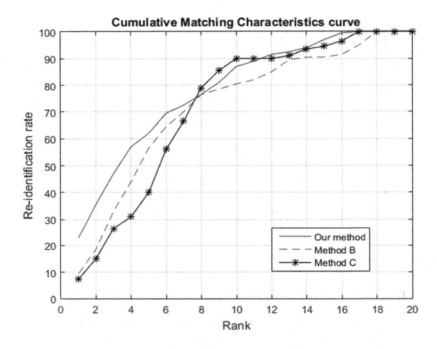

**Fig. 6.** CMC curve for selected images evaluation.

**Table 3.** Re-identification performance for selected images evaluation.

|  | Rank-1 | Rank-2 | Rank-3 | Rank-10 | nAUC |
|---|---|---|---|---|---|
| Our method | **23** | **35.5** | **47** | 87 | **79.63** |
| Method B | 9.5 | 18.5 | 33 | 80.5 | 73.72 |
| Method C | 7.5 | 15 | 26.5 | **90** | 73.61 |

We can see that the performance of our method is much better than that of method B and method C in the first ranks of the CMC curve and the nAUC. However, the major drawback of the randomly selected images is they may have a redundant information or uncompleted visual details that will be considered in the global appearance signature. Also, they perform uniformly for each target and does not take in consideration the unpredictable target's path. On the other hand, our method automatically selects a set of representative appearance images based on the silhouette's contour variations from the target's trajectory, operating on each target independently. In addition, no learning phase is needed that makes it suitable for real scenario.

## 4    Conclusion

In this paper, we have presented a novel unsupervised appearance key frame selection approach for multi-shot person re-identification problem. More precisely, we have proposed to automatically select a small set of representative appearance images depicting the different body postures by analyzing the silhouette's contour variations from the target's trajectory. Our approach does not need any training data which is a huge advantage for real applications. The visual characteristics of the selected images are modeled into a global appearance signature, and, the similarity are computed based a the average of all pairwise distances. We evaluated our method on the challenging HDA+ dataset. We have proved robustness of the proposed approach against state-of-the-art approaches.

Encouraged by the promising performance of our method, we are currently examining how to improve the results by grouping the selected appearance image in different semantic clusters based on the estimation of the body's pose. Images with the same pose are likely to be compared in the matching step. Such step has an important impact on further stages such reduce the temporal and spatial complexity of the re-identification problem. In addition, we aim to evaluate the proposed approach on other person re-identification datasets.

**Acknowledgment.** This work was supported by the PHC Utique program for the CMCU DEFI project (N 34882WK).

## References

1. Doretto, G., Sebastian, T., Tu, P.H., Rittscher, J.: Appearance-based person reidentification in camera networks: problem overview and current approaches. J. Ambient Intell. Humaniz. Comput. **2**, 127–151 (2011)
2. Bird, N.D., Masoud, O., Papanikolopoulos, N.P., Isaacs, A.: Detection of loitering individuals in public transportation areas. IEEE Trans. Intell. Transp. Syst. **6**(2), 167–177 (2005)
3. Cheng, D.S., Cristani, M., Stoppa, M., Bazzani, L., Murino, V.: Custom pictorial structures for re-identification. In: British Machine Vision Conference, pp. 1–11 (2011)
4. Zeng, M., Wu, Z., Tian, C., Zhang, L., Hu, L.: Efficient person re-identification by hybrid spatiogram and covariance descriptor. In: IEEE Conference on Computer Vision and Pattern Recognition Workshops, pp. 48–56 (2015)
5. Vezzani, R., Baltieri, D., Cucchiara, R.: People reidentification in surveillance and forensics: A survey. ACM Comput. Surv. **46**(2), 29.1–29.37 (2013)
6. Bazzani, L., Cristani, M., Perina, A., Farenzena, M., Murino, V.: Multiple-shot person re-identification by hpe signature. In: 20th International Conference on Pattern Recognition (2010)

7. Farenzena, M., Bazzani, L., Perina, A., Murino, V., Cristani, M.: Person reidentification by symmetry-driven accumulation of local features. In: IEEE Conference on Computer Vision and Pattern Recognition (2010)
8. Bazzani, L., Cristani, M., Murino, V.: SDALF: modeling human appearance with symmetry-driven accumulation of local features. In: Gong, S., Cristani, M., Yan, S., et al. (eds.) Person Re-Identification, pp. 43–69. Springer, London (2014)
9. Bak, S., Zaidenberg, S., Boulay, B., Bremond, F.: Improving person reidentification by viewpoint cues. In: 11th IEEE International Conference on Advanced Video and Signal-based Surveillance, pp. 175–180 (2014)
10. Wang, T., Gong, S., Zhu, X., Wang, S.: Person re-identification by video ranking. In: 13th European Conference Computer Vision (2014)
11. Figueiredo, M., Jain, A.: Unsupervised learning of finite mixture models. IEEE Trans. Pattern Anal. Mach. Intell. **24**(3), 381–396 (2002)
12. Dalal, N., Triggs, B.: Histograms of oriented gradients for human detection. In: IEEE Conference on Computer Vision and Pattern Recognition, pp. 886–893 (2005)
13. Frikha, M., Fendri, E., Hammami, M.: A new appearance signature for real time person re-identification. In: 15th International Conference Intelligent Data Engineering and Automated Learning, pp. 175–182 (2014)
14. Pudil, P., Novovicova, J., Kittler, J.: Floating search methods in feature selection. In: 12th International Conference on Pattern Recognition (1994)
15. Figueira, D., Taiana, M., Nambiar, A., Nascimento, J., Bernardino, A.: The HDA+ data set for research on fully automated re-identification systems. In: Workshop of European Conference on Computer Vision, pp. 241–255 (2014)
16. Bhattacharyya, A.: On a measure of divergence between two statistical populations defined by their probability distribution. Bull. Calcutta. Math. Soc. **35**, 99–109 (1943)
17. Buchsbaum, G.: A spatial processor model for object colour perception. J. Franklin Institute **300**, 1–26 (1980)
18. Prosser, B., Zheng, W.S., Gong, S., Xiang, T.: Person re-identification by support vector ranking. In: British Machine Vision Conference (2010)
19. Gray, D., Tao, H.: Viewpoint invariant pedestrian recognition with an ensemble of localized features. In: 10th European Conference on Computer Vision, pp. 262–275 (2008)
20. Layne, R., Hospedales, T.M., Gong, S.: Attributes-based re-identification. In: Gong, S., et al. (eds.) Person Re-Identification, pp. 93–117. Springer, Heidelberg (2014)
21. Martinel, N., Micheloni, C., Foresti, G.L.: A pool of multiple person re-identification experts. Pattern Recogn. Lett. **71**, 23–30 (2016)
22. Liu, J., Liang, C., Ye, M., Wang, Z., Yang, Y., Han, Z., Sun, K.: Person re-identification via attribute confidence and saliency. In: 16th Pacific-Rim Conference on Multimedia, pp. 591–600 (2015)
23. Wang, T., Gong, S., Zhu, X., Wang, S.: Person re-identification by discriminative selection in video ranking. IEEE Trans. Pattern Anal. Mach. Intell. **38**, 2501–2514 (2016)

# An Exact Smoother in a Fuzzy Jump Markov Switching Model

Zied Bouyahia[1(✉)], Stéphane Derrode[2], and Wojciech Pieczynski[3]

[1] Department of Computer Science, CAAS, Dhofar University, Salalah, Oman
bouyahiazied@gmail.com
[2] Ecole Centrale de Lyon, LIRIS, CNRS UMR 5205, Écully, France
stephane.derrode@ec-lyon.fr
[3] Telecom Sudparis, SAMOVAR, CNRS UMR 5157, Évry, France
wojciech.pieczynski@telecom-sudparis.eu

**Abstract.** In this paper, we proposed an extension of the classical Conditionally Gaussian Observed Markov Switching Model (CGOMSM) by incorporating fuzzy switches. The proposed approach allows the modeling of transient switches and handles the discontinuity feature in switching regime models by using fuzzy switches instead of hard jumps. Fuzzy switched based approach is more adapted to real-world application in which regime continuity is an intrinsic property. To define an efficient scheme for an exact smoothing in CGOMFSM, we adapt fast smoothing equations to cope with the fuzzy model. Finally, we show through several experiments the interest of the fuzzy switches model.

**Keywords:** Fuzzy switching models · Exact smoothing · Non-linear Markov systems

## 1 Introduction

Let $\mathbf{X}_1^N = \{X_1, \ldots, X_N\}$, $\mathbf{Y}_1^N = \{Y_1, \ldots, Y_N\}$ and $\mathbf{R}_1^N = \{R_1, \ldots, R_N\}$ be three random sequences taking values in $\mathbb{R}^m$, $\mathbb{R}^q$ and $\Omega = \{1, \ldots, K\}$ respectively. Let $\mathbf{X}_1^N$ be a hidden process and $\mathbf{Y}_1^N$ be an observed process. We consider a switching regime model represented by the sequence of switches $\mathbf{R}_1^N$. We address the smoothing problem consisting in an recursive search of the unobserved process $\mathbf{X}_1^N$ and the switches sequence $\mathbf{R}_1^N$, only knowing the observed sequence $\mathbf{Y}_1^N$. A fast Bayesian processing can be carried out by assuming that the distribution of $(\mathbf{X}_1^N, \mathbf{Y}_1^N)$ is within the framework of hidden Gaussian Markov model. The non-linearity can be modeled by a switching regime system. Then, the idea is to approximate a non-linear non-Gaussian system by a regime switching Gaussian system. Some recent switching models have been proposed with efficient fast exact filtering schemes. These switching models called "conditionally Markov switching hidden linear models" (CMSHLM) [16] include conditionally Gaussian observed Markov switching models (CGOMSM) defined as follows:

© Springer International Publishing AG 2017
B. Ben Amor et al. (Eds.): RFMI 2016, CCIS 684, pp. 111–125, 2017.
DOI: 10.1007/978-3-319-60654-5_10

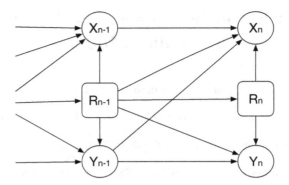

**Fig. 1.** Directed graph representing dependencies between random sequences $\mathbf{X}_1^N$ $\mathbf{Y}_1^N$ and $\mathbf{R}_1^N$. Circles represent continuous process and diamond represents discrete process.

- $T_1^N = (X_1^N, R_1^N, Y_1^N)$ is Markov chain;
- $p(r_{n+1}|x_n, r_n, y_n) = p(r_{n+1}|r_n)$ and

$$
\begin{bmatrix} X_{n+1} \\ Y_{n+1} \end{bmatrix} = \begin{bmatrix} A_{n+1}^{xx}(R_n^{n+1}) & A_{n+1}^{xy}(R_n^{n+1}) \\ 0 & A_{n+1}^{yy}(R_n^{n+1}) \end{bmatrix} \begin{bmatrix} X_n \\ Y_n \end{bmatrix}
$$
$$
+ \begin{bmatrix} B_{n+1}^{xx}(R_n^{n+1}) & B_{n+1}^{xy}(R_n^{n+1}) \\ B_{n+1}^{yx}(R_n^{n+1}) & B_{n+1}^{yy}(R_n^{n+1}) \end{bmatrix} \begin{bmatrix} U_{n+1} \\ V_{n+1} \end{bmatrix} + \begin{bmatrix} N^X(R_n^{n+1}) \\ N^Y(R_n^{n+1}) \end{bmatrix},
$$

with

$$
\begin{bmatrix} N^X(R_n^{n+1}) \\ N^Y(R_n^{n+1}) \end{bmatrix} = \begin{bmatrix} M^X(R_{n+1}) - A_{n+1}^{xx}(R_n^{n+1})M^X(R_n) - A_{n+1}^{xy}(R_n^{n+1})M^Y(R_n) \\ M^Y(R_{n+1}) - A_{n+1}^{yy}(R_n^{n+1})M^Y(R_n) \end{bmatrix},
$$

where $U_1^N$ and $V_1^N$ are two Gaussian unit-variance white noise vectors, $M^X(R_n)$ and $M^Y(R_n)$ are respective means of $\mathbf{X}_1^N$ and $\mathbf{Y}_1^N$ in each state (independently from $n$). The CGOMSM is then defined by matrices $A(\mathbf{R}_n^{n+1})$, $B(\mathbf{R}_n^{n+1})$, transition matrix denoted as $t$ such that $t(i, j) = p(r_{n+1} = j|r_n = i)$ and mean vectors $M(R_n) = [M^X(R_n); M^Y(R_n)]$. Figure 1 depicts the dependence graph of the CGOMSM model.

We assume that $\mathbf{R}_1^N$ takes its values in a discrete finite set of $K$ switches $\Omega = \{1, ..., K\}$. This hard jumps model has been widely used in several contexts dealing with switching regime Markov systems. Its success comes from its ability to represent non-linear dynamic patterns which is an inherent property in several applications (analysis of economic and finance time series [11], sustainable energy [6], robotics [7,8,12], etc.).

However, this model does not take into account the intrinsic imprecision of the switches in real-world applications. In fact, hard jumps induce discontinuity in the dynamic behavior of the studied system. This transitory imprecision can be handled with fuzzy modeling which consists in allowing each switch to take its value as a mixture of many components simultaneously. Fuzzy modeling has been widely incorporated in several applications dealing with Markov models [13–15]. In this paper, we present a new method to approximate non-linear Markov systems using a new variant of CGOMSM using fuzzy switches (hereafter called CGOMFSM).

The remaining of this paper is organized as follows. In the second section, we detail the formulation of fuzzy switching model. The third section describes the adaptation of CGOMSM algorithms for parameters estimation and for posterior marginal probabilities computation the fuzzy counterpart. The fourth section presents experimental results, and the last one draws conclusions and future work.

## 2   Fuzzy Switching Model with $K$ Hard Classes

### 2.1   Probability Distribution of Fuzzy Vectors

In the fuzzy switches system which extends the hard case with $K$ classes $\Omega = \{1, ..., K\}$, we assume that each jump $r_n^K$ is a vector in $[0,1]^K$. So $r_n^K = (\varepsilon_1, ..., \varepsilon_K)$, and each component $\varepsilon_k$ can be seen as "fuzzy part" of the hard class $k$. Therefore, we have $\sum_{k=1}^{K} \varepsilon_k = 1$. This is an extension of the hard case as each $\varepsilon = (\varepsilon_1, ..., \varepsilon_K)$ of the form $\varepsilon = (0, ..., 0, \varepsilon_k = 1, 0, ..., 0)$ is assimilated to the hard class $k$. The distribution of $R_n^K$ is then a probability distribution on $F^K \subset [0,1]^K$, with $F^K$ the set of $\varepsilon = (\varepsilon_1, ..., \varepsilon_K)$ verifying $\sum_{k=1}^{K} \varepsilon_k = 1$. Such a probability distribution can be defined in different manners; we propose to adopt the following one. Let us consider $\delta_0$ Dirac mass on 0 and $\delta_1$ Dirac mass on 1, and let $\mu$ be Lebesgue measure on $]0,1[$. Let us note $\nu = \delta_0 + \delta_1 + \mu$. Since $\varepsilon_K = 1 - (\varepsilon_1 + ... + \varepsilon_{K-1})$, it is sufficient to define the distribution of the random vector $R_n^{K-1} = (R_n^1, ..., E_n^{K-1})$ on $F_{K-1} \subset [0,1]^{K-1}$ whose elements verify $\varepsilon_1 + ... + \varepsilon_{K-1} \leq 1$. Let us define this distribution by a density $f$ with respect to the measure $\nu^{\otimes(K-1)}$. Thus, we have:

$$P_{R^{K-1}} = f\nu^{\otimes(K-1)}. \tag{1}$$

The case $K = 2$ will be dealt with in next sections, in the example below we specify the case $K = 3$.

*Example.* Let $K = 3$. Here the distribution on $F_2 \subset [0,1]^2$ (whose elements verify $\varepsilon_1 + \varepsilon_2 \leq 1$) is $P_{E^2} = f\nu^{\otimes 2} = f(\delta_0 + \delta_1 + \mu)^{\otimes 2} = f(\delta_{00} + \delta_{01} + \delta_{10} + \delta_{11} + \delta_0 \otimes \mu + \mu \otimes \delta_0 + \delta_1 \otimes \mu + \mu \otimes \delta_1 + \mu \otimes \mu)$.

However, we recall that $f(\varepsilon_1, \varepsilon_2) = 0$, for $(\varepsilon_1, \varepsilon_2) \in [0,1]^2$ such that $\varepsilon_1 + \varepsilon_2 > 1$. The sets $F^3 \subset [0,1]^3$ and $F_2 \subset [0,1]^2$ are illustrated in Fig. 2.

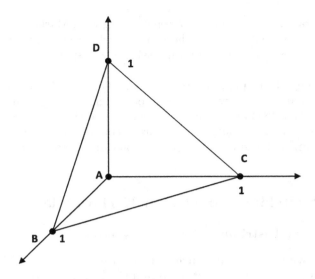

**Fig. 2.** For three hard classes, domain $F^3$ is the triangle DB, while domain $F_2$ is the triangle ABC.

## 2.2 Joint Densities

In Markov context considered in this paper, we have to define joint densities of the distributions of $(R_n, R_{n+1})$ defined on $F_{K-1} \times F_{K-1}$. We will consider $f$ of the form:

$$f(r_n, r_{n+1}) = \begin{cases} \alpha_{ij} & \text{if both switches are hard,} \\ \beta\phi(r_n, r_{n+1}) + \theta & \text{otherwise.} \end{cases} \quad (2)$$

We choose $\phi(r_n, r_{n+1}) = \left(1 - \frac{\|r_{n+1}-r_n\|}{\sqrt{2}}\right)^r$, $r \in \mathbb{R}$ with $\|r_{n+1} - r_n\|$ is the distance between two consecutive switches given by the quadratic norm. Then:

$$\phi(r_n, r_{n+1}) = \left(1 - \left(\frac{\sum_{i=1}^{K} (\varepsilon_n^i - \varepsilon_{n+1}^i)^2}{2}\right)^{\frac{1}{2}}\right)^r, \quad r \in \mathbb{R}. \quad (3)$$

*Example.* When the number of hard switches equals 3, the expression of normalization condition gives:

$$\int_0^1 \int_0^1 f(u,v)(\delta_0 + \delta_1 + \mu)^{\otimes 2}(u,v) =$$

$$\alpha_{00} + \alpha_{01} + \alpha_{10} + \alpha_{11} + \beta \left[ \int_0^1 \phi(0,v)dv + \int_0^1 \phi(u,0)du + \int_0^1 \int_0^1 \phi(u,v)dudv \right] + \theta = 1,$$

$$(4)$$

with $\phi(u,v) = (1 - |u - v|)^r$.

### 2.3   Parameters Interpolation

The model matrices of the fuzzy switching model can be calculated by linear interpolation using the following formula:

$$A(\varepsilon_1, \varepsilon_2) = \sum_{1 \le i,j \le K} \varepsilon_1^i \varepsilon_2^j A(i,j),$$

$$B(\varepsilon_1, \varepsilon_2) = \sum_{1 \le i,j \le K} \varepsilon_1^i \varepsilon_2^j B(i,j),$$

$$M(\varepsilon_1, \varepsilon_2) = \sum_{1 \le i,j \le K} \varepsilon_1^i \varepsilon_2^j M(i,j),$$

where $A(i,j)$ and $B(i,j)$ are the model matrices corresponding to the hard components $i$ and $j$. $M(i,j)$ is the mean vector for hard switches $i$ and $j$.

The implementation of the fuzzy switching model can be performed by an adequate quantification of the interval $[0,1]$ into $F$ discrete fuzzy levels. The larger $F$ is, the more accurate the representation of data would be. However, choosing a large number of fuzzy levels will lead to high computation time. For example, when the number of crisp components equals three, setting $F = 3$ yields 15 switches and setting $F = 4$ gives 21 switches.

## 3   Fuzzy Switching Model with Two Hard Components

In this remaining of the paper, we consider the case of two hard switches $\Omega = \{0,1\}$. To model fuzzy switches, we consider that each random variable $R_n$ in $\mathbf{R}_1^N$ takes its values in the continuous interval $[0,1]$, instead of the set $\{0,1\}$.

Let us denote the pair $(\varepsilon_n^0, \varepsilon_n^1) \in [0,1]$, in which $\varepsilon_n^i$ represents the contribution of the hard component $i$ to the switch $r_n$. Without loss of generality, let $\varepsilon_n = \varepsilon_n^1 = 1 - \varepsilon_n^0$. Then we have $R_n = \varepsilon_n$:

- $\varepsilon_n = 0$ if the switch is the hard component 0.
- $\varepsilon_n \in ]0,1[$ if the switch is fuzzy.
- $\varepsilon_n = 1$ if the switch is the hard component 1.

So this model is able to represent signals with both discrete (hard) and continuous (fuzzy) components. Let $\nu = \delta_0 + \delta_1 + \mu$.

We will assume that $\mathbf{R}_1^N$ is stationary. Then $p(r_n^{n+1})$ does not depend on $n$. Let us now precisely define the joint a priori density $p(r_1^2)$, where notation $r_1^2$ represents the pair $(r_1, r_2)$. $p(r_1^2)$ is defined with respect to the measure product $\nu \otimes \nu$, under normalization condition (Fig. 3).

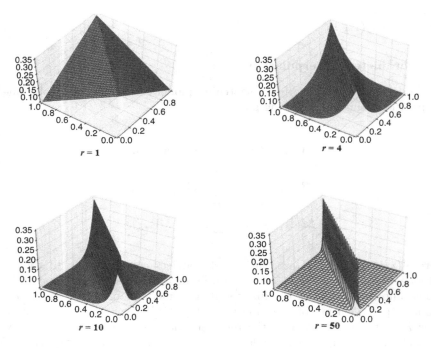

**Fig. 3.** Density of $p\left(r_1^2\right)$ with respect to measure $\nu \otimes \nu$.

Analytic computation of the joint prior densities can be worked out by quantifying the interval $[0, 1]$ into $F$ equal-length sub-intervals $\left[\frac{i}{F}, \frac{i+1}{F}\right]$ as described in Fig. 4. Using this scheme, the normalization condition in Eq. (4) yields:

$$\alpha_{00} + \alpha_{01} + \alpha_{10} + \alpha_{11}$$
$$+ \beta \left[ \frac{1}{2F} \sum_{i=0}^{F-1} (1 - \varepsilon_i)^r + \frac{1}{2F} \sum_{i=0}^{F-1} \varepsilon_i^r + \frac{1}{2F^2} \sum_{i=0}^{F-1} \sum_{j=0}^{F-1} (1 - |\varepsilon_i - \varepsilon_j|)^r \right] + \theta = 1. \quad (5)$$

Each sub-interval can be represented by its medium value $\frac{2i+1}{2F}$. So, in this discrete approximate scheme, the joint a priori density can be defined by a $(2 + F) \times (2 + F)$ matrix.

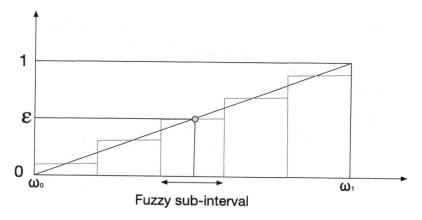

**Fig. 4.** Subdivision of the interval $[0, 1]$ into $F = 5$ equal-length fuzzy sub-intervals.

Under the assumptions of fuzzy switches, we can define the matrices of the incorporated model using a bi-linear function as follows:

$$A(\varepsilon_1, \varepsilon_2) = [(1 - \varepsilon_1)A(0,0) + \varepsilon_1 A(1,0)](1 - \varepsilon_2)$$
$$+ [(1 - \varepsilon_1)A(0,1) + \varepsilon_1 A(1,1)]\varepsilon_2 \qquad (6)$$

$$B(\varepsilon_1, \varepsilon_2) = [(1 - \varepsilon_1)B(0,0) + \varepsilon_1 B(1,0)](1 - \varepsilon_2)$$
$$+ [(1 - \varepsilon_1)B(0,1) + \varepsilon_1 B(1,1)]\varepsilon_2 \qquad (7)$$

The means vectors of the fuzzy model are calculated using the following equations:

$$M(\varepsilon_i) = [(1 - \varepsilon_i)M(0) + \varepsilon_i M(1)] \qquad (8)$$

Hence, the CGOMFSM is entirely defined by

- the parameters of the corresponding deterministic hard switching model,
- the number of fuzzy levels $F$, and
- parameters $\alpha_{ij}, i, j \in \{0, 1\}$, $\beta$, $\theta$ and $r$.

The parameter $r$ specifies the homogeneity of the switching model. The larger $r$ is, the larger the probability of having two similar consecutive switches is.

Figure 5 represents an example of simulation of $(\mathbf{X}_1^N, \mathbf{Y}_1^N, \mathbf{R}_1^N)$ using the set of parameters of a fuzzy switching model defined in Table 1. Simulations were performed using the following transition matrix:

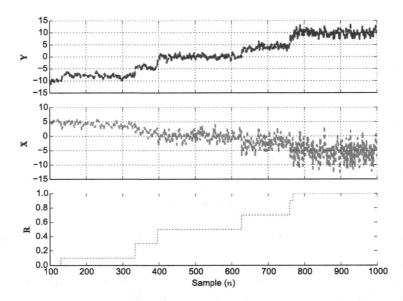

**Fig. 5.** Trajectories of simulated CGOMFSM $(\mathbf{X}_1^N, \mathbf{Y}_1^N, \mathbf{R}_1^N)$.

$$
t = \begin{pmatrix}
0.99 & 0.01 & 0 & 0 & 0 & 0 & 0 \\
0 & 0.99 & 0.01 & 0 & 0 & 0 & 0 \\
0 & 0 & 0.99 & 0.01 & 0 & 0 & 0 \\
0 & 0 & 0 & 0.99 & 0.01 & 0 & 0 \\
0 & 0 & 0 & 0 & 0.99 & 0.01 & 0 \\
0 & 0 & 0 & 0 & 0 & 0.99 & 0.01 \\
0 & 0 & 0 & 0 & 0 & 0 & 1.00
\end{pmatrix}.
$$

In the transition matrix rows correspond to $R_n$ and columns correspond to $R_{n+1}$. The first and the last rows and columns represent hard switches while other rows and columns correspond to discrete fuzzy switches. This simulation shows the imprecision between hard switches 0 and 1 as illustrated in the trajectory of $\mathbf{Y}_1^N$. The choice of the transition matrix allows a progressive regime switching of parameters sets corresponding to hard switch 0 to hard switch 1.

## 4    Fast Smoothing in the CGOMFSM

Let us denote by $\mathbf{T}_1^N$ the triplet $(\mathbf{X}_1^N, \mathbf{R}_1^N, \mathbf{Y}_1^N)$. The smoothing problem consists in computing:

$$
\mathbb{E}\left[\mathbf{X}_{n+1} \big| \mathbf{y}_1^N\right] = \int_{[0,1]} p\left(r_{n+1} = \nu \big| \mathbf{y}_1^N\right) \mathbb{E}\left[\mathbf{X}_{n+1} \big| r_{n+1} = \nu, \mathbf{y}_1^{n+1}\right] d\nu, \quad (9)
$$

**Table 1.** Example of fuzzy switching model with 5 fuzzy switches.

| $(R_n, R_{n+1})$ | $(0,0)$ | $(1,1)$ |
|---|---|---|
| $(MX, MY)$ | $(-5, 10)$ | $(5, -10)$ |
| $A$ | 0.55 0.35<br>0.00 0.83 | 0.40 0.20<br>0.00 0.46 |
| $B$ | 0.51 0.00<br>0.28 0.48 | 0.86 0.00<br>0.17 0.87 |

from $p\left(r_{n+1} \mid \mathbf{y}_1^N\right)$ and $\mathbb{E}\left[\mathbf{X}_{n+1} \mid r_{n+1}, \mathbf{y}_1^{n+1}\right]$.

The optimal smoother computes recursively $p\left(r_{n+1} \mid \mathbf{y}_1^N\right)$ and $\mathbb{E}\left[\mathbf{X}_{n+1} \mid r_{n+1}, \mathbf{y}_1^{n+1}\right]$ from $p\left(r_n \mid \mathbf{y}_1^N\right)$ and $\mathbb{E}\left[\mathbf{X}_n \mid r_n, \mathbf{y}_1^n\right]$ and the model parameters using the procedure detailed in [9]. The main difference between CGOMSM and CGOMFSM is that, in the case of fuzzy switches, we involve continuous integration, requiring to be quantified with respect to the number of discrete fuzzy levels $F$.

Since $(\mathbf{R}_1^N, \mathbf{Y}_1^N)$ is a pairwise Markov chain in the model, we get

$$p\left(r_{n+1} \mid \mathbf{y}_1^{n+1}\right) = \frac{\int_{[0,1]} p\left(r_{n+1}, \mathbf{y}_{n+1} \mid r_n = \nu, \mathbf{y}_n\right) p\left(r_n = \nu \mid \mathbf{y}_1^n\right) d\nu}{\int_{[0,1]} \int_{[0,1]} p(r_{n+1}^* = v, \mathbf{y}_{n+1} \mid r_n = \nu, \mathbf{y}_n) p\left(r_n = \nu \mid \mathbf{y}_1^n\right) d\nu dv}, \quad (10)$$

and

$$p\left(r_n \mid r_{n+1}, \mathbf{y}_1^{n+1}\right) = \frac{\int_{[0,1]} p\left(r_{n+1}, \mathbf{y}_{n+1} \mid r_n, \mathbf{y}_n\right) p\left(r_n \mid \mathbf{y}_1^n\right) d\nu}{\int_{[0,1]} p\left(r_{n+1}, \mathbf{y}_{n+1} \mid r_n^*, \mathbf{y}_n\right) p\left(r_n^* \mid \mathbf{y}_1^n\right) d\nu}. \quad (11)$$

Since

$$\mathbb{E}\left[\mathbf{X}_n \mid \mathbf{r}_n^{n+1}, \mathbf{y}_1^{n+1}\right] = \mathbb{E}\left[\mathbf{X}_n \mid r_n, \mathbf{y}_1^n\right], \quad (12)$$

and from (9), we can derive the following recursive equation:

$$\mathbb{E}\left[\mathbf{X}_{n+1} \mid r_{n+1}, \mathbf{y}_1^{n+1}\right] = \int_{[0,1]} p\left(r_n \mid r_{n+1}, \mathbf{y}_1^{n+1}\right) \times$$

$$F_{n+1}(\mathbf{r}_n^{n+1}, \mathbf{y}_n^{n+1}) \mathbb{E}\left[\mathbf{X}_n \mid r_n, \mathbf{y}_1^n\right] + H_{n+1}(\mathbf{r}_n^{n+1}, \mathbf{y}_n^{n+1}) d\nu, \quad (13)$$

with $F_{n+1}(\mathbf{r}_n^{n+1}, \mathbf{y}_n^{n+1})$ and $H_{n+1}(\mathbf{r}_n^{n+1}, \mathbf{y}_n^{n+1})$ are adequate matrices. Probabilities $p(r_n \mid y_1^n)$ and $p(r_n, y_1^N)$ are recursively calculated in linear time using forward and backward probabilities in the Markov chain $(Y_1^N, R_1^N)$ such that $\alpha_n(r_n) = p(r_n, y_1^n)$ and $\beta_n(r_n) = p(y_{n+1}^N \mid r_n, y_n)$.

$$\alpha_1(r_1) = p(r_1, y_1)$$
$$\alpha_{n+1}(r_{n+1}) = \int_{[0,1]} \alpha_n(v) p(r_{n+1}, \mathbf{y}_{n+1} \mid r_n, \mathbf{y}_n) dv \quad (14)$$

and

$$\beta_N(r_N) = 1$$

$$\beta_n(r_n) = \int_{[0,1]} \beta_{n+1}(v)p(r_{n+1}, \mathbf{y}_{n+1}|r_n, \mathbf{y}_n)dv \tag{15}$$

Using forward-backward probabilities, we can compute the smoothed and the filtered probabilities as follows:

$$p(r_n|\mathbf{y}_1^N) = \frac{\alpha_n(r_n)\beta_n(r_n)}{\int_{[0,1]} \alpha_n(v)\beta_n(v)dv} \tag{16}$$

$$p(r_n|\mathbf{y}_1^n) = \frac{\alpha_n(r_n)}{\int_{[0,1]} \alpha_n(v)dv}. \tag{17}$$

Posterior marginal probabilities are calculated using the normalized Baum-Welch algorithm. The algorithm computes recursively the forward and backward probabilities. In the case of fuzzy switches, these probabilities are defined as follows:

$$\alpha_{n+1}(\delta) = \int_{[0,1]} \alpha_n(\theta)p(\mathbf{t}_{n+1}(\delta)|\mathbf{t}_n(\theta))\,d\theta, \tag{18}$$

$$\beta_n(\delta) = \int_{[0,1]} \beta_{n+1}(\theta)p(\mathbf{t}_{n+1}(\delta)|\mathbf{t}_n(\theta))\,d\theta, \tag{19}$$

with $\mathbf{t}_n(\theta) = (\mathbf{x}_n, \mathbf{y}_n, r_n = \theta)$.
Then:

$$p(r_n, r_{n+1}|\mathbf{x}_1^N, \mathbf{y}_1^N) = \frac{\alpha_n(r_n)p(\mathbf{t}_{n+1}|\mathbf{t}_n)\beta_{n+1}(r_{n+1})}{\int_{[0,1]} \int_{[0,1]} \alpha_n(\delta)p(\mathbf{t}_{n+1}|\mathbf{t}_n)\beta_{n+1}(\theta)d\delta d\theta}. \tag{20}$$

## 5    Experiments

In this section, we present two series of experiments to assess the performance of the exact smoother, in the case of scalar data ($m = q = 1$). In the first series we evaluate the performance of the fuzzy model with synthetic fuzzy signals; in the second series we apply our algorithm to smooth simulated Stochastic Volatility (SV) data. In both experiments, parameters estimation is carried out using EM algorithm using training samples denoted by $(\mathbf{x}_1^T, \mathbf{y}_1^T)$ of size $T$. Then we repeatedly generate, according to the considered model, synthetic sequences of size $S$ denoted by $(\mathbf{x}_1^S, \mathbf{y}_1^S)$. Smoothing algorithm is then performed using estimated parameters to generate $\hat{\mathbf{x}}_1^S$ from the observed sequence $\mathbf{y}_1^S$. The criterion used to assess the efficiency of smoothing algorithms is the mean squared error (MSE) defined as follows:

$$MSE = \frac{1}{S} \sum_{n=1}^{S} (x_n - \hat{x}_n)^2 \tag{21}$$

## 5.1  Smoothing Synthetic Fuzzy Signals

We define the distribution of the random process $(\mathbf{X}_1^N, \mathbf{Y}_1^N, \mathbf{R}_1^N)$ by the Gaussian distributions $p(z_1, z_2 | r_1, r_2)$, where $Z_n = (X_n^T, Y_n^T)^T$. Let $\Gamma_{i,j}$ be the covariance matrix $\Gamma_{i,j} = \mathbb{E}\left[ Z_1 Z_2^T | r_1 = i, r_2 = j \right] = \begin{bmatrix} 1 & b_i & a_{ij} & d_{ij} \\ b_i & 1 & e_{ij} & c_{ij} \\ a_{ij} & e_{ij} & 1 & b_j \\ d_{ij} & c_{ij} & b_j & 1 \end{bmatrix}$.

To ensure that the model follows the definition of CGOMSM ($A^{yx}(R_n^{n+1}) = 0$), we take $d_{ij} = b_i c_{ij}$, we also take $a_{ij} = c_{ij}$ and $e_{ij} = d_{ij}$. Moreover, we consider that $a_{ij} = a_i$. Hence, the simulation model is defined by the set of parameters $\Theta = \{a_0, a_1, b_0, b_1\}$. We consider five cases defined by the parameter set $\Theta$ detailed in Table 2. For each case, we generate a fuzzy signal with 5 discrete fuzzy switches. Then we restore the hidden process using the CGOMFSM model using different values of $F$ ranging from 1 to 5. For each set of data, we consider 3 values of $r \in \{2, 8, 20\}$. For each case, we perform 10 independent experiments with $S = 1000$. Each experiment consists in generating a training sample $(\mathbf{X}_1^T, \mathbf{Y}_1^T)$ of size $T = 20000$ for the 100 iterations of EM algorithm. Figure 6 shows examples of trajectories of simulated process $\mathbf{R}_1^N$ with three different values of $r$. Figure 7 illustrates an example of simulated data using case 2 parameters and the optimal (but approximated) smoothing output. Table 3 reports the MSE results for the 5 different cases of fuzzy models and different values of $r$.

**Table 2.** Parameters of 5 simulation cases.

|        | Case1 | Case 2 | Case 3 | Case 4 | Case 5 |
|--------|-------|--------|--------|--------|--------|
| $a_0$  | 0.4   | 0.6    | 0.8    | 0.9    | 0.1    |
| $a_1$  | 0.5   | 0.2    | 0.4    | 0.5    | 0.7    |
| $b_0$  | 0.3   | 0.3    | 0.4    | 0.4    | 0.7    |
| $b_1$  | 0.1   | 0.2    | 0.3    | 0.4    | 0.5    |

The experimental results show that when the number of fuzzy levels increases, the smoothed signal is closer to the "ground-truth" hidden signal. Moreover, we noticed that the higher the homogeneity parameter $r$, the lower the estimation error.

## 5.2  Experiments on Stochastic Volatility Models

Stochastic volatility (SV) models are widely used to highlight the variance of stochastic processes [10]. Several variants of SV models have been studied

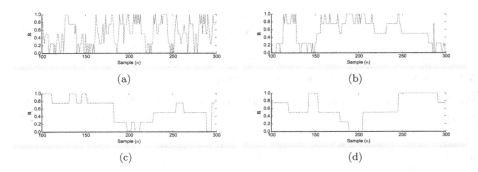

**Fig. 6.** Examples of trajectories of simulated process $\mathbf{R}_1^N$ using different values of switches homogeneity $r = 2$ (a), $r = 8$ (b), $r = 20$ (c) and $r = 40$ (d).

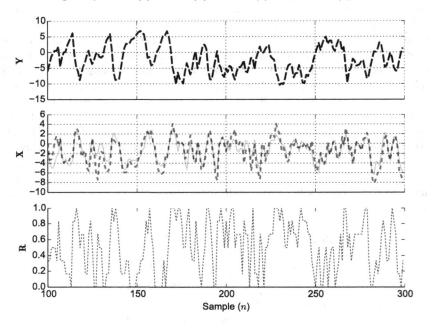

**Fig. 7.** A $(\mathbf{X}_1^N, \mathbf{R}_1^N, \mathbf{Y}_1^N)$ CGOMFSM trajectory together with the restored signal in green (solid). (Color figure online)

(Henston, CEV, GARCH, Chen, etc.). In this paper, we consider the standard SV model defined as follows:

$$X_1 = \mu + U_1 \tag{22}$$

$$X_{n+1} = \mu + \phi(X_n - \mu) + \sigma U_{n+1} \tag{23}$$

$$Y_n = \beta \exp\left(\frac{X_n}{2}\right) V_n, \tag{24}$$

**Table 3.** MSE results for different fuzzy signals with 5 discrete fuzzy levels and different values of $r$.

| $F$ | $r$ | Case 1 | Case 2 | Case 3 | Case 4 | Case 5 |
|---|---|---|---|---|---|---|
| 0 | 2 | 1.117 | 1.155 | 1.130 | 1.217 | 1.491 |
|   | 8 | 0.993 | 1.033 | 1.011 | 1.111 | 1.217 |
|   | 20 | 0.884 | 0.897 | 0.888 | 0.923 | 0.988 |
| 1 | 2 | 0.942 | 1.004 | 1.023 | 1.173 | 1.240 |
|   | 8 | 0.804 | 0.829 | 0.84 | 0.944 | 1.035 |
|   | 20 | 0.722 | 0.724 | 0.733 | 0.808 | 0.912 |
| 2 | 2 | 0.815 | 0.803 | 0.887 | 0.858 | 1.008 |
|   | 8 | 0.732 | 0.736 | 0.770 | 0.798 | 0.88 |
|   | 20 | 0.683 | 0.697 | 0.701 | 0.762 | 0.804 |
| 3 | 2 | 0.738 | 0.701 | 0.705 | 0.781 | 0.923 |
|   | 8 | 0.653 | 0.648 | 0.641 | 0.709 | 0.813 |
|   | 20 | 0.603 | 0.618 | 0.603 | 0.666 | 0.747 |
| 5 | 2 | 0.721 | 0.691 | 0.688 | 0.732 | 0.852 |
|   | 8 | 0.644 | 0.635 | 0.626 | 0.662 | 0.763 |
|   | 20 | 0.598 | 0.602 | 0.589 | 0.621 | 0.711 |

where $U_i$, $V_i$ are independent standard Gaussian vectors. The SV models is defined by the set of parameters $\sigma$, $\mu$, and $\alpha$. The main conclusion is that when the number of discrete fuzzy states increases, the model approaches the results of the optimal (but time consuming) particle smoother (Table 4).

# 6   Conclusion

In this paper, we presented a novel approach to approximate non-linear Markov system using Conditionally Gaussian Observed Markov Fuzzy Switching Model (CGOMFSM). The chief novelty of this work is the introduction of fuzzy jumps instead of classical crisp states. This model still allows exact (up to required

**Table 4.** MSE for SV models with $\mu = 0.5, \beta = 0.5$. To ensure stationarity of models, we set $\phi^2 = 1 - \sigma^2$. PS column is the result obtained from the particle smoother (1500 particles).

|  |  | $F$ |  |  |  |  |
|---|---|---|---|---|---|---|
|  | $\phi$ | $\sigma$ | 0 | 1 | 3 | 5 | PS |
| Case 1 | 0.99 | 0.141 | 0.395 | 0.233 | 0.144 | 0.124 | 0.12 |
| Case 2 | 0.9 | 0.435 | 0.480 | 0.387 | 0.350 | 0.339 | 0.33 |
| Case 3 | 0.8 | 0.6 | 0.557 | 0.496 | 0.474 | 0.467 | 0.46 |
| Case 4 | 0.5 | 0.866 | 0.701 | 0.672 | 0.661 | 0.661 | 0.66 |

quantification) and fast smoothing equations. The fuzzy jumps allow transient modification of parameters, which is more appropriate for real-world applications. Future work includes the evaluation of the model on real data.

# References

1. Abbassi, N., Benboudjema, D., Derrode, S., Pieczynski, W.: Optimal filter approximations in conditionally gaussian pairwise Markov switching models. IEEE Trans. Autom. Control **60**(4), 1104–1109 (2015)
2. Salzenstein, F., Collet, C.: Fuzzy Markov random fields versus chains for multispectral image segmentation. IEEE Trans. Pattern Anal. Mach. Intell. **28**(11), 1753–1767 (2006)
3. Caillol, H., Hillion, A., Pieczynski, W.: Fuzzy random fields and unsupervised image segmentation. IEEE Trans. Geosci. Remote Sensing **34**(4), 801–810 (1993)
4. Gorynin, I., Derrode, S., Monfrini, E., Pieczynski, W.: Exact fast smoothing in switching models with application to stochastic volatility. In: EUSIPCO, Nice, France, pp. 924–928, 31 August - 4, September 2015
5. Caillol, H., Pieczynski, W., Hillon, A.: Estimation of fuzzy Gaussian mixture and unsupervised statistical image segmentation. IEEE Trans. Image Process. **6**(3), 425–440 (1997)
6. Yang, L., He, M., Zhang, J., Vittal, V.: Support vector machine enhanced Markov model for short term wind power forecast. IEEE Trans. Sustain. Energy **6**(3), 791–799 (2015)
7. Artemiadis, P.K., Kyriakopoulos, K.J.: A switching regime model for the EMG-based control of a robot arm. IEEE Trans. Syst. Man Cybern. Part B (Cybernetics) **41**(1), 53–63 (2011)
8. Corff, S.L., Fort, G., Moulines, E.: Online expectation maximization algorithm to solve the SLAM problem. In: 2011 IEEE Statistical Signal Processing Workshop (SSP), pp. 225–228, Nice, June 2011
9. Gorynin, I., Derrode, S., Monfrini, E., Pieczynski, W.: Fast filtering in switching approximations of non-linear Markov switching systems with application to stochastic volatility. IEEE Trans. Autom. Control **62**(2), 853–862 (2017)
10. Ghysels, E., Harvey, A., Renault, E.: Stochastic volatility. In: Handbook of Statistics, vol. 14, pp. 119–192 (1995)
11. Koko, M.: Application of Markov-switching model to stock returns analysis. Dyn. Econometric Models **7**, 259–268 (2006)
12. Baltzakis, H., Trahanias, P.: A hybrid framework for mobile robot localization: formulation using switching state-space models. Auton. Robots **15**(2), 169–191 (2003)
13. Salzenstein, F., Pieczynski, W.: Parameter estimation in hidden fuzzy Markov random fields and image segmentation. Graph. Model Image Process. **59**(4), 205–220 (1997)
14. Carincotte, C., Derrode, S., Sicot, G., Boucher, J.M.: Unsupervised Image segmentation based on a new fuzzy hidden Markov chain model. In: IEEE International Conference on Acoustic, Speech, Signal Processing, Montreal, Canada, May 2004 (2004)
15. Carincotte, C., Derrode, S., Bourennane, S.: Unsupervised change detection on SAR images using fuzzy hidden Markov chains. IEEE Trans. Geosci. Remote Sensing **44**(2), 432–441 (2006)

16. Pieczynski, W.: Exact filtering in conditionally Markov switching hidden linear models. Comptes Rendus Mathématique **349**(9–10), 587–590 (2011)
17. Gamal-Eldin, A., Salzenstein, F., Collet, C.: Hidden fuzzy Markov chain model with K discrete classes. In: 10th International Conference on Information Science, Signal Processing and their Applications (ISSPA 2010), Kuala Lumpur, pp. 109–112, May 2010

# 2D Shape Analysis

# A Novel 2D Contour Description Generalized Curvature Scale Space

Ameni Benkhlifa[✉] and Faouzi Ghorbel[✉]

Cristal Laboratory, Grift Group, National School of Computer Sciences,
Université de la Manouba, 2010 Manouba, Tunisia
ameni.benkhlifa@ensi-uma.tn, faouzi.ghorbel@ensi.rnu.tn
http://www.ensi.rnu.tn/

**Abstract.** Here, we intend to propose a 2D contour descriptor that
we call Generalized Curvature Scale Space (GCSS) based on the iso-
curvature levels, and the curvature scale space (CSS) descriptor. We start
by computing the curvature in different scales and extract the points
which have the same curvature values as the maximums in each scale.
Each CSS image is represented by a set of key points. The Dynamic Time
Warping (DTW) similarity measure is used. We reach a significant rate
in image recognition using two data sets (HMM GPD and MPEG7 CE
Shape-1 Part-B set).

**Keywords:** Pattern recognition · Contour 2D · Curvature ·
Iso-curvature · CSS

## 1 Introduction

Henceforth, the intervention of the human being is insubstantial for several applications. Mentioning as examples, the segmentation of tumors and organs from
medical data that have become more and more automatic, face recognition, especially with the increase of the number of crimes and terrorism, robot navigation
etc. Thus, it is obvious to pass by a shape recognition step. However, it is not an
easy task for many reasons. The object may undergo geometric transformations
or nonlinear deformations like noise and occlusion. To overcome those problems,
the object must be described by a discriminant, efficient, stable and robust shape
descriptor which is invariant with respect to any transformation belongs to the
planar Euclidean transformations group $E(2)$.

In this part, we present an overview of shape representation and description
techniques. The existing shape analysis methods can be classified into three fam-
ilies: contour-based, region-based and hybrid methods that combine the contour
and region based methods.

Among the first category, the Fourier descriptors [2–4] which focuses on the
global features of the curve. Thus, other methods treat local features such as
Hoffman and Richards [6] who partitioned the curve into parts at negative curvature minima which enhanced the object recognition. An other work in the

© Springer International Publishing AG 2017
B. Ben Amor et al. (Eds.): RFMI 2016, CCIS 684, pp. 129–140, 2017.
DOI: 10.1007/978-3-319-60654-5_11

same context was proposed by Siddiqi and Kimia [7] based on the segmentation of the silhouette. Xu [8] proposed a new method called contour flexibility which represents the deformable potential at each point along a contour. Klassen et al. [9] presented a differential geometric curve representation using its direction functions and curvature functions. Shu [10] proposed a descriptor named contour points distribution histogram which is based on the distribution of points on object contour under polar coordinates. An other contour-based method proposed by Mokhtarian [1] is the curvature scale space which is based on the computation of the maxima of the curvature of the shape smoothed by Gaussian filters in different scales.

In the region-based family, we find methods who are based on 2D Zernike moments such as [11]. Those methods are very sensitive to local changes such as occlusion or overlapping objects. Analytical Fourier Mellin transform gives also invariant descriptors for region [5]. An other method is proposed by Hong [12] based on a kernel descriptor that characterizes local shape. There are also the third category which is the hybrid methods that combine both contour-based and region-based methods together. A descriptor called Rolling penetrate [13] which uses a set of scanning lines that rotate around the shape centroid to collect information.

We propose here a novel contour-based descriptor, called Generalized Curvature Scale Space (**GCSS**). Our approach is based jointly on the iso-curvature parameterization which is invariant under Euclidean transformations and CSS [1] method that determines the extremums of the contour in different scales. For each scale, we extract the local extremums of the curvature. We select only those superior to a given threshold. Our descriptor is formed by all curve points having the selected curvature levels. These key points, extracted from different scales. GCSS gives an $E(2)$ invariant non uniform parameterization of the contour since it redistributes the points in the regions that have strong curvature. We use the Dynamic Time Wraping as a similarity metric for the shape recognition.

The following paper is organized as follows: we describe the steps of our approach in the second section. In the third section, we expose the used similarity metric that corresponds to the Dynamic Time Warping. In the fourth section we show and discuss the results of the application of our approach using the MPEG7 dataset and HMM GPD dataset.

## 2    Generalized Curvature Scale Space (GCSS)

We aim in this section to describe the 2D contour by a robust, accurate and invariant descriptor.

We start by extracting the boundary of the 2D shape of a binary image and we obtain a set of points which are different from a shape to another.

### 2.1    Parameterization Using the Arc Length

Assuming that the boundary coordinates of the 2D shape is a closed contour. Let C be a curve and its parameterization $C(t)$ is a function of a parametric variable $t$ defined as:

$$C : [0,1] \rightarrow \mathbb{R}^2$$
$$t \mapsto [x(t), y(t)]^t \qquad (1)$$

Since the parameterization of the curve is not unique, because it depends on the starting point, the orientation of the points and the speed we go over the curve, it is difficult to compare between contours properly. A solution for this problem is to use the arc length reparamerization of the contour. It is well known that it is invariant under Euclidean transformations which conserve the Euclidean norm, the length, the inner product and the speed we go over the curve.

$$S(t) = 1/L \int \sqrt{x_t(u)^2 + y_t(u)^2} du, t \in [0,1] \qquad (2)$$

Where $L$ denotes the length of the contour $C$.

## 2.2 Parameterization Using the Uniform Iso-curvature

The uniform iso-curvature is a parameterization of the contour $C$. The idea of this parameterization is to find the points of iso-curvature which are a set of uniform levels of the curvature values.

After reparameterization of the contour $C$ with the arc length, since it reduces the expression of the curvature, the number of points in each shape boundary is the same. The curvature $\kappa$ of the shape boundary points is defined as follows:

$$\kappa(s) = \frac{x_s(s)y_{ss}(s) - y_s(s)x_{ss}(s)}{(x_s^2(s) + y_s^2(s))^{3/2}} \qquad (3)$$

Where $x_s(s)$, $x_{ss}(s)$, $y_s(s)$ and $y_{ss}(s)$ are respectively the first and the second derivatives of $x(s)$ and $y(s)$. The curvature describes well the speed of turning a contour.

After normalizing $\kappa$ in $[0, 1]$, it will be divided in a finite number of the uniform levels (equidistant levels) $\ell_{uni}$.

$$\hat{\kappa}(s) = |\frac{\kappa(s)}{max|\kappa(s)|}| \qquad (4)$$

where $\hat{\kappa}$ is the normalized curvature. After chosen the uniform iso-curvature points, we obtain the following set of unordered points of $C$:

$$\mathbf{L} = \{(x(s_i^j), y(s_i^j)); \quad s_i^j = \hat{\kappa}^{-1}\{\ell_{uni}^j\}\} \qquad (5)$$

Where $\hat{\kappa}^{-1}$ is the inverse normalized curvature function.

We propose to order the set $\mathbf{L}$ by going over the curve in a given sense.

The major problem of this approach is the choice of the adequate number of levels to describe well the contour and if we increase infinitly the number of levels any point of the curve can be seen as limit of a sequence points from $\mathbf{L}$. Figure 1 shows that we obtain different iso-curvature points by applying different numbers of uniform levels on the same shape.

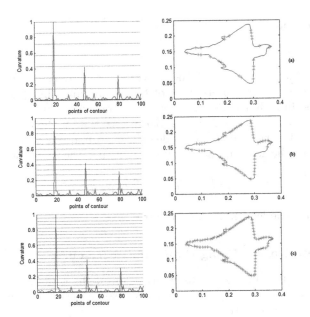

**Fig. 1.** Isocurvature points for 3 different number of uniform levels for the same shape (a) 8 levels (b) 16 levels (c) 24 levels

## 2.3   Curvature Scale Space

The curvature scale space (CSS) [1] is a shape representation for planar curves based on finding the extremums of the curve in different scales by computing the convolution of the shape boundary with Gaussian kernels.

$$\begin{cases} x(s,\sigma) = x(s) \otimes g(s,\sigma) \\ y(s,\sigma) = y(s) \otimes g(s,\sigma) \end{cases} \tag{6}$$

Where $g(s,\sigma)$ is a Gaussian function and $\otimes$ is the convolution product.
   The curvature $\kappa(s,\sigma)$ at each point of the curve is:

$$\kappa(s,\sigma) = \frac{x_s(s,\sigma)y_{ss}(s,\sigma) - y_s(s,\sigma)x_{ss}(s,\sigma)}{(x_s^2(s,\sigma) + y_s^2(s,\sigma))^{3/2}} \tag{7}$$

Where $x_s(s,\sigma)$, $y_s(s,\sigma)$, $x_{ss}(s,\sigma)$ and $y_{ss}(s,\sigma)$ are respectively the first and the second derivatives of $x(s,\sigma)$ and $y(s,\sigma)$ and they are given by:

$$\begin{cases} x_s(s,\sigma) = x(s) \otimes g_s(s,\sigma) & x_{ss}(s,\sigma) = x(s) \otimes g_{ss}(s,\sigma) \\ y_s(s,\sigma) = y(s) \otimes g_s(s,\sigma) & y_{ss}(s,\sigma) = y(s) \otimes g_{ss}(s,\sigma) \end{cases} \tag{8}$$

   Let $C_\sigma$ be the curve $C$ smoothed with $g(s,\sigma)$. Thus, as we increase the value of $\sigma$, $C_\sigma$ becomes smoother and the number of extremums decreases more and more. For example, in Fig. 2 we test four different $\sigma$: 1, 2, 5 and 10 from the left to the right.

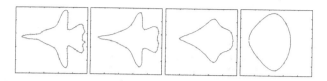

**Fig. 2.** CSS images for the same shape (plane) in four different scales: 1, 2, 5 and 10 from the left

## 2.4   GCSS Computing

In this paragraph, we will describe our proposed descriptor. It is obtained by combining the CSS [1] and the iso-curvatures parameterization. Our iso-curvatures levels are the curvature variations of the extremums of the contour in given scales. Thus our levels are not equidistant.

The steps of our approach are:

1. First of all, we find the extremums $\ell_\sigma$ of the smoothed shape boundary $C_\sigma$.
2. We threshold $\ell_\sigma$ to avoid the zero curvature points. We obtain our nonuniform iso-curvature levels $\ell_\sigma(\tau)$.

$$\ell_\sigma(\tau) = \{\ell_\sigma; \qquad \ell_\sigma > \tau\} \tag{9}$$

3. Then, $F(\sigma)$ is our descriptor. It contains the curvature values of intersection points of $\hat{\kappa}_\sigma$ (the normalized curvature $\hat{\kappa}(s,\sigma)$ is replaced by $\hat{\kappa}_\sigma$) with $\ell_\sigma(\tau)$. $F_{coor}(\sigma)$ represents the coordinates of the key points found in $F(\sigma)$.

$$F(\sigma) = \{\hat{\kappa}_\sigma(s_i^j); \qquad s_i^j = \hat{\kappa}_\sigma^{-1}\{\ell_\sigma^j(\tau)\}\} \tag{10}$$

$$F_{coor}(\sigma) = \{(x(s_i^j,\sigma), y(s_i^j,\sigma)); \qquad s_i^j = \hat{\kappa}_\sigma^{-1}\{\ell_\sigma^j(\tau)\}\} \tag{11}$$

$F_{coor}(\sigma)$ contains the coordinates of the key points that have the same curvature values as the extremums of $\hat{\kappa}_\sigma$ for the curve $C_\sigma$ in the scale $\sigma$.

For a finite number of scales, we obtain our descriptor $F$ and its corresponding $F_{coor}$ described as follows:

$$F = \bigcup_{\sigma \in \Sigma} F(\sigma) \tag{12}$$

$$F_{coor} = \bigcup_{\sigma \in \Sigma} F_{coor}(\sigma) \tag{13}$$

We compute $F$ and $F_{coor}$ in a very restrict set of scales $\Sigma$.

We explain in Fig. 3 the steps of GCSS in just one scale ($\sigma = 5$).

The GCSS descriptor presents also a new parameterization of the contour. Such parameterization is invariant as the curvature. However, in the discret case, $F_{coor}$ can be seen as a resampling procedure which gives us an unordered set of

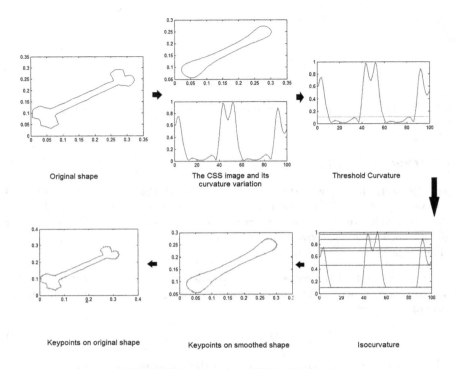

Fig. 3. The proposed GCSS with $\sigma = 5$

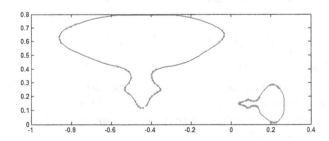

Fig. 4. Invariance of GCSS to geometric transformations

points of $C$. In the last task, the obtained set of points is ordered. Such set is distributed non uniformly and we have more points in strong curvature. Figure 4 demonstrates well the distribution of the key points and their invariance under E(2). However, it depends on the points permutation. The use of dynamic time Warping as distance between descriptors achieves the invariance relatively to the permutations.

Although CSS [1] provides the extremums of the smoothed shape boundary, it ignores the neighbors of those points because it represents the local curve by only one point which is the local extremum. In an other hand, a lot of points could be selected in localities having low curvatures (less geometrical information). Figures 5 and 6 illustrate well the weakness of CSS in front of GCSS in different scales.

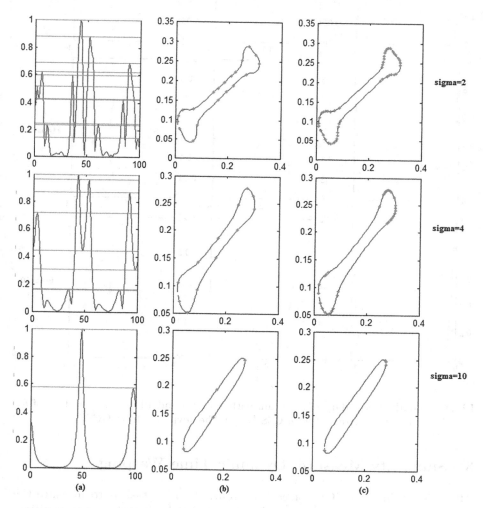

**Fig. 5.** Smoothed boundary of Bone with $\sigma = 2, 4$ and 10 (a) the variation of $\hat{\kappa}_\sigma$ (b) The extremums extracted from CSS [1] (c) The keypoints from GCSS

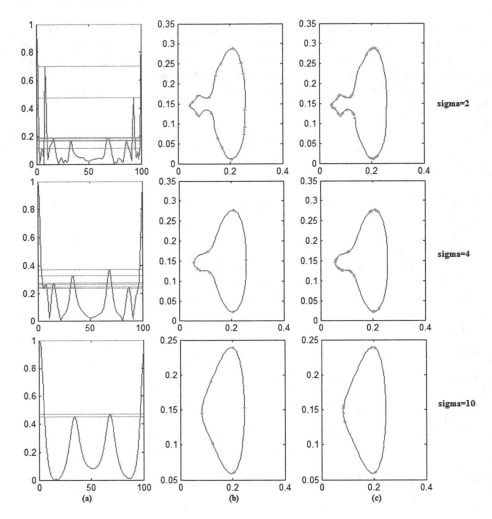

**Fig. 6.** Smoothed boundary of Fountain with $\sigma = 2, 4$ and 10 (a) the variation of $\hat{\kappa}_\sigma$ (b) The extremums extracted from CSS [1] (c) The keypoints from GCSS

## 3    Similarity Measure: Dynamic Time Warping

After computing the GCSS of each contour, the next task is to measure the similarity or the distance between descriptors that leads to better recognition. Since the length of GCSS vector is different from a contour to an other, we use the dynamic time warping (DTW) algorithm as it was introduced in [14]. The dynamic time warping does not depend on the vector length.

For two shapes $A$ and $B$ we obtain respectively $F(A)$ and $F(B)$. The distance between two series $F(A) = \{F(a_i); \ i = 1 : N\}$ and $F(B) = \{F(b_j); \ j = 1 : M\}$, where $N$ and $M$ are respectively the cardinalities of $F(A)$ and $F(B)$, is defined by:

$$D(F(a_i), F(b_j)) = min \left\{ \begin{array}{l} D(F(a_i), F(b_{j-1})) \\ D(F(a_{i-1}), F(b_j)) \\ D(F(a_{i-1}), F(b_{j-1})) \end{array} \right. + D(F(a_i), F(b_j)) \qquad (14)$$

## 4   Experiments and Results

The performance of our novel descriptor Generalized Curvature Scale Space is tested. Two data sets are used for the experimentations: HMM GPD and MPEG7 CE Shape-1 Part-B data set [15]. For the recognition, each object is compared to all the shapes in the data set using the DTW algorithm and matched to the closest one.

### 4.1   The Data Sets

The HMM GPD is composed by four sub data sets: bicego-data [16] is formed by 140 objects grouped in 7 classes and 20 elements, plane-data [17] contains 210 objects (7 classes and 30 elements), mpeg-data [17] and car-data [17] have 120 objects where mpeg is grouped in 6 classes and 20 elements and the car data set has 4 classes and 30 elements. We form a data set using the four sub data sets (bicego, plane, car and mpeg) by picking up the 20 first elements of each classes. Thus we obtain 480 objects regrouped in 24 classes and 20 objects in each class.

The MPEG7 CE Shape-1 Part-B data set [15] is a well known data set. It is composed of 1400 elements that are grouped in 70 classes and 20 images in each class. Figure 7 shows some examples of each data set.

**Fig. 7.** Data sets used in the experimentations: left MPEG7 data set; right HMMGPD data set

## 4.2  Results

We choose two scales which are 5 and 6 to compute GCSS since they give a significant set of points, and we fixed $\tau = 10^{-3}$ to eliminate the points that have very low curvature. For the recognition step, the pairwise shape matching scores using the knearest neighbor (kNN) algorithm is used. In this algorithm, for each shape, the distance DTW is computed with all the other shapes in the data set and we search the knearest neighbor.

We compare our proposed descriptor GCSS to CSS [1] with $\sigma$ ranges from 1 to 10 with the step = 0.1. Table 1 lists the retrieval results of our descriptor GCSS on HMM GPD data set using 1NN algorithm. We reach very high score for Mpeg (99.16%) and Plane (98.57%) data sets. For MPEG7 CE Shape-1 Part-B [15], Fig. 8 shows that we reach 89.43% using the kNN algorithm for the 5 nearest neighbors, meanwhile CSS [1] reaches only 78%.

The proposed GCSS outperforms CSS [1] for both data sets HMM GPD and MPEG7 CE Shape-1 Part-B [15].

Figure 9 illustrates the variation of the retrieval results on Bicego data set as a function of the parameter $\tau$. The retrieval rate reaches 95.7% when $10^{-3} < \tau < 10^{-2}$.

**Table 1.** Retrieval results on HMM data set using 1NN algorithm for: GCSS $\sigma \in [5, 6]$ and $\tau = 10^{-3}$ and CSS [1]

|        | Rate GCSS (%) | Rate CSS (%) |
|--------|---------------|--------------|
| Bicego | 95.71         | 90.00        |
| Plane  | 98.57         | 79.52        |
| Car    | 73.33         | 55.00        |
| Mpeg   | 99.16         | 95.83        |
| HMM    | 88.12         | 75.62        |

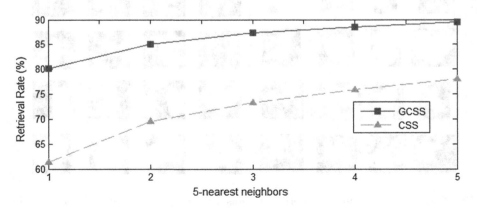

**Fig. 8.** Retrieval Rate for MPEG7 CE Shape-1 Part-B data set for: GCSS ($\sigma \in [5, 6]$ and $\tau = 10^{-3}$) and CSS ($\sigma \in [1:10]$ with the step = 0.1)

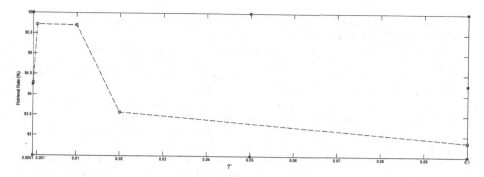

**Fig. 9.** The impact of the choice of $\tau$ on the retrieval rate for Bicego data set; $\tau = [10^{-4}, 10^{-3}, 10^{-2}, 0.02, 0.1]$

## 5   Conclusion

We have presented a new shape boundary descriptor that we named Generalized Curvature Scale Space. The GCSS is a set of points that have the same curvature of the maximas of the contour in different scales. Our descriptor reaches significant rates by testing on two data sets HMM GPD and MPEG7 CE SHAPE-1 Part-B comparing to CSS [1].

The GCSS represents also a new parameterization for the contour since it describes it well in the regions that have strong curvature. It is invariant under $E(2)$ group.

Several future works are possible for our descriptor. We try to make the choice of the parameter of GCSS, such as $\sigma$ and $\tau$, adaptative to the shape boundary. GCSS also can be combined with other shape representations.

## References

1. Mokhtarian, F., Abbasi, S., Kittler, J.: Robust and efficient shape indexing through curvature scale space. In: Proceedings of the 1996 British Machine and Vision Conference BMVC, vol. 96 (1996)
2. Wallace, T.P., Wintz, P.A.: An efficient three-dimensional aircraft recognition algorithm using normalized Fourier descriptors. Comput. Graph. Image Process. **13**(2), 99–126 (1980)
3. Persoon, E., King-Sun, F.: Shape discrimination using Fourier descriptors. IEEE Trans. Syst. Man. Cybern. **7**(3), 170–179 (1977)
4. Ghorbel F.: Stability of invariants Fourrier descriptors and its interference on the shape classification. In: 11th International Conference on Pattern Recognition, (ICPR), 30 August–3 September, The Hague (1992)
5. Ghorbel, F.: Towards a unitary formulation for invariant image description: application to image coding. Annales des telecommunications **53**(5–6), 242 (1998). Springer
6. Hoffman, D.D., Richards, W.A.: Parts of recognition. Cognition **18**(1), 65–96 (1984)

7. Siddiqi, K., Kimia, B.B.: Parts of visual form: computational aspects. IEEE Trans. Pattern Anal. Mach. Intell. **17**(3), 239–251 (1995)
8. Shu, X., Xiao-Jun, W.: A novel contour descriptor for 2D shape matching and its application to image retrieval. Image Vis. Comput. **29**(4), 286–294 (2011)
9. Xu, C., Liu, J., Tang, X.: 2D shape matching by contour flexibility. IEEE Trans. Pattern Anal. Mach. Intell. **31**(1), 180–186 (2009)
10. Klassen, E., et al.: Analysis of planar shapes using geodesic paths on shape spaces. IEEE Trans. Pattern Anal. Mach. Intell. **26**(3), 372–383 (2004)
11. Khotanzad, A., Hong, Y.H.: Invariant image recognition by Zernike moments. IEEE Trans. Pattern Anal. Mach. Intell. **12**(5), 489–497 (1990)
12. Hong, B.-W., et al.: Shape representation based on integral kernels: application to image matching and segmentation. In: 2006 IEEE Computer Society Conference on Computer Vision and Pattern Recognition (CVPR 2006), vol. 1. IEEE (2006)
13. Chen, Y.W., Xu, C.L.: Rolling penetrate descriptor for shape-based image retrieval and object recognition. Pattern Recognit. Lett. **30**(9), 799–804 (2009)
14. Sankoff, D., Kruskal, J.B.: Time warps, string edits, and macromolecules: the theory and practice of sequence comparison. In: David, S., Kruskal, J.B. (eds.), vol. 1. Addison-Wesley Publication, Reading (1983)
15. Latecki, L.J., Lakamper, R., Eckhardt, T.: Shape descriptors for non-rigid shapes with a single closed contour. In: Proceedings of the IEEE Conference on Computer Vision and Pattern Recognition, vol. 1. IEEE (2000)
16. Bicego, M., Murino, V., Figueiredo, M.A.T.: Similarity-based classification of sequences using hidden Markov models. Pattern Recognit. **37**(12), 2281–2291 (2004)
17. Thakoor, N., Gao, J., Jung, S.: Hidden Markov model-based weighted likelihood discriminant for 2-D shape classification. IEEE Trans. Image Process. **16**(11), 2707–2719 (2007)

# A Coronary Artery Segmentation Method Based on Graph Cuts and MultiScale Analysis

Chaima Oueslati[1,2]([✉]), Sabra Mabrouk[1,2], Faouzi Ghorbel[1,2],
and Mohamed Hedi Bedoui[1,2]

[1] CRISTAL Laboratory, GRIFT Research Group,
National School of Computer Science, 2010 Manouba, Tunisia
{chaima.oueslati,sabra.mabrouk}@ensi-uma.tn, faouzi.ghorbel@ensi.rnu.tn,
medhedi.bedoui@fmm.rnu.tn
[2] Laboratory of Biophysics, Faculty of Medicine of Monastir, TIM Team,
University of Monastir, 5019 Monastir, Tunisia

**Abstract.** In this paper we propose a new multi-scale fully automatic
algorithm based on Graph cuts for vessel extraction. In fact, we combine
vesselness, geodesic paths, a multi-scale edgeness map and the directional
information for vessel tracking in order to personalize the Graph cuts
approach to the segmentation of tubular structures.

**Keywords:** X-Ray · Angiography · Segmentation · Graph cuts

## 1 Introduction

Cardiovascular diseases are the leading cause of death in the developed countries
essentially due to coronary atherosclerosis [1]. The medical imaging modality
and the minor procedures most currently used to diagnosis the coronary diseases are the X-ray angiography. As angiographic images are subject to noise
and radiological contrast agent that is widely heterogeneous, it is difficult to
identify the arteries compared to the background. Segmentation is therefore necessary to extract the coronary arteries by eliminating artifacts contained in the
background. Several methods were proposed for the coronary artery segmentation. In fact, Vessel segmentation algorithms are the fundamental component of
automated radiological diagnostic systems such as diagnosis of the vessels (e.g.
stenosis or malformations) and registration of patient images obtained at different times. Two categories of vessel segmentation algorithms are distinguished.
The first one is skeleton based technique which aim first to extract median
blood vessel and then connect these centerlines and estimate the vessel width
to develop the vessel tree. The main idea of non skeleton methods is to directly
extract blood vessels based mostly to the pixels intensity. In this context, several works were proposed to solve the vessel segmentation issue including basic
ones such as thresholding and morphological operator. In [2], the authors propose a fuzzy clustering where each data point can belong to more than one
cluster and clusters are determined via similarity measures, such as distance,

B. Ben Amor et al. (Eds.): RFMI 2016, CCIS 684, pp. 141–151, 2017.
DOI: 10.1007/978-3-319-60654-5_12

connectivity, or intensity. [3,4] use thresholding to extract vessels from angiographic images using the intensity information to assign pixels into categories. Mathematical morphology was proposed in [5], the authors provide systematic approach to analyze geometric Characteristic of artery by applying structuring elements :Dilation and erosion (SE) to images. In [6] the authors present a region growing procedure that segment images by incrementally selecting pixels into the artery region based on similarity and spatial proximity criteria. In [7], authors define a new segmentation method using an iteration region growing and the Sato vesselness function by merging both vesselness and direction information for vessel tracking. This method was extended in [8] where authors change sato filter with Frangi filter because of the sensitivity of the Sato filter which is higher than the Frangi filter [9].

More sophisticated techniques were also used to detect vessels such as deformable models which find the vessel contours using parametric curves that deform under the influence of internal and external forces [10].

In [14], the authors propose a method for segmenting coronary arteries based on graph cuts using multi-scale vesselness measure. In [11], the authors adapt the graph cut method for segmentation of thin blood vessels since the original method suffers from the shrinking bias drawback [12] which makes it not suitable for segmentation of thin objects like blood vessels.

In the latter work, the authors offer an energy functional GC which takes into account the probability of tabular structures, local connectivity of the vessel region and measuring edgeness.

In this paper we propose a new multi-scale Graph cut method for coronary artery segmentation in 2D X-ray angiograms by combining Vesselness measure to visualize the local vessel, geodesic paths to emphasize the local connectivity of another vessel region, a multi-scale edgeness map using the adaptive Canny detector and the directional information to guide the segmentation of blood vessels. The remainder of the paper is organized as follows: In Sect. 2, we remind the Graph cut method, Sect. 3 describes our proposed method. Experiments are presented and discussed in Sect. 4. And finally we conclude the paper by giving some details on future work.

## 2    Graph Cut Method

In this section, we overview graph cut (GC) method presented in [11]. The main idea of this latter paper is to use an energy formulation which takes into account the local vessel appearance using a vesselness measure, the local connectivity to other vessel regions using geodesic paths and a measure of edgeness based on a multi-scale adaptive Canny detector.

Image is seen as non-oriented graph where V is the set of nodes corresponding to the pixels and L is the set of arcs referring to the adjacency relations between pixels with two additional nodes the source s and the sink t. The graph arcs are composed of:

- s-links: arcs connecting a node i to the source s;
- t-links: arcs connecting a node i to the sink t;
- n-links: arcs connecting neighbor pixels i and j.

A cut is a partition $\{S, T\}$ of nodes into two disjoint sets S and T where the source s belongs to S and the sink t belongs to T. The aim of the method is to find the optimal cut C with the lowest cost. The cost of the cut can be defined as:

$$C(S, T) = \sum_{i \in S,\ j \in T} C(i, j). \tag{1}$$

Where $C(i, j)$ is the capacity of the arc connecting nodes i and j.

The t-links and s-links between a pixel i and the two nodes S and T terminals are weighted by a term named unary term U related to regional properties of the segmentation. This term can be considered as the probability that the pixel i belongs to the class $\omega$ (i.e. $U_i(\text{"vess"})$ or $U_i(\text{"back"})$). Every pixel is connected with its N-neighborhood (N = 4,8) by a segment called n-link. The n-links are weighted by a regularization term designed to ensure the spatial coherence in a pixel neighborhood named boundary term $B\{i, j\}$.

Any $B\{i, j\} \geq 0$ can be interpreted as the presence of a continuity between i and j. The total energy of a Cut C in the graph is defined by:

$$E(C) = \sum_{i \in V} U_i(w) + \sum_{i \in N, j \in N} B\{i, j\} \times \delta(i, j). \tag{2}$$

Where V is the set of vertices (nodes), $\delta(i, j) = 0$ if $\omega_i = \omega_j$ and 1, otherwise.

In [11] authors define the unary terms by combining vesselness map $V(i)$ and the geodesic distance map Di, we note that the measure of vesselness is not totally effective in the detection of tabular structures since there are some vessel regions, which can have low vessel probabilities, so authors defeated this limitation by introducing a vesselness-geodesic measure VG which employ the geodesic paths among vessel seeds and compute geodesic distance among vessel seeds. In fact, the geodesic distance map includes the distance of each pixel to a set of centroid of a k-means clustering on the Cartesian coordinates of the vessels seeds. The map VG for pixel i is defined by the authors as:

$$VG_i = max\left(V(i)\ , \frac{\left(\frac{1}{(D_i + \mu(D))}\right)}{\left(max\left(\frac{1}{(D + \mu(D))}\right)\right)}\right) \tag{3}$$

where $D$ correspond to the geodesic distance map and $\mu(D)$ is its mean and $V(i)$ is the measure of vesselness for the pixel i.

This measure $V(i, \sigma)$ with i: indice, $\sigma$: scale is calculated as follows:

$$V(i, \sigma) = \begin{cases} 0 & if\ \lambda_2 > 0 \\ \exp\left(\frac{-R_\beta^2}{2 \times \beta^2}\right) \times \left(1 - exp\left(\frac{-A^2}{2 \times C^2}\right)\right) & otherwise \end{cases} \tag{4}$$

where $R_\beta = \frac{\lambda_1}{\lambda_2}$, $\Delta = \lambda_1^2 + \lambda_2^2$, $\beta$ and c are thresholds which control the filter's sensitivity to $R_\beta$ and $\Delta$ respectively, $\lambda_1(i,\sigma)$ and $\lambda_2(i,\sigma)$ where $|\lambda_1(i,\sigma)| <= |\lambda_2(i,\sigma)|$ are the eigenvalues of the Hessian matrix of image I computed at scale $\sigma$ and location i.

$R_\beta$ is the blob-like structure measure, this measure is used to distinguish between structures which form lines and flat structures.

$\Delta$ is the Frobenius norm of the Hessian matrix, this value is small in the background with individually small eigenvalues by lack of contrast.

This Eq. (4) can be evaluated at different scales Q, $\sigma = \{\sigma^1,..,\sigma^Q\}$, and the V for each pixel is given by $V(i) = max_{\sigma \in \{\sigma^1,..,\sigma^Q\}} V(i,\sigma)$.

The unary terms for each pixel i can therefore be written as:

$$U_i (\text{"vess"}) = -\ln(p(\omega = \text{"vess"})) \tag{5}$$

$$U_i (\text{"back"}) = -\ln(p(\omega = \text{"back"})) \tag{6}$$

Where $p(\omega = \text{"vess"}) = VG_i$ is the probability of a pixel belonging to a vessel region, $p(\omega = \text{"back"}) = 1 - p(\omega = \text{"vess"})$ is the opposite probability.

In order to initialize the boundary term, authors use the Canny edge detector algorithm on the image at different threshold levels.

## 3    The Proposed Method

In this work, we propose a GC new energy functional adapted to the segmentation of vessel problem taking into consideration direction information given by the first Hessian eigenvector in order to extract thin vessels in addition to the vesselness measure, the geodesic path and the multi-scale edgeness measure. The direction and the normal direction of the potential linear structure describe the local shape of the intensity and can be determined respectively by computing the eigenvectors $e_1$ and $e_2$ of the hessian matrix. The direction information $D_\sigma$ of a pixel i is given by the following expression:

$$D_\sigma(i) = \begin{cases} 0 \; if \; \lambda_2 < 0 \\ e_1 \; otherwise. \end{cases} \tag{7}$$

Where $\lambda_2$ is the eigen value of the Hessian matrix of image I computed at scale $\sigma$ and location i.

The unary term is computed identically to the described graph cut method by combining the vesselness map and the geodesic map. For boundary terms, the key idea is to compute them in such a way they guarantee the continuity of the segmentation along the artery direction.

The initial boundary potential over the multiscale edgeness map is computed as $B\{i,j\} = J_i^*$.

$J_i^*$ is computed as follows:

$$J_i^* = min_f \frac{1}{n} \sum_{e=1}^{n} J_{i,\theta_e,\alpha^f} \tag{8}$$

where $\theta_e$ is the threshold, $\alpha^f$ is a scale and $J_{i,\theta_e,\alpha^f}$ is the binary edge map for pixel i.

Scales should be adjusted between the approximate width of the smallest and largest vessel to be detected. For different scales $\sigma$, the direction $(D_\sigma)$ is computed and the measure $B\{i,j\}$ is increased only if the pixels i and j have close directions even if their the edge probability values are lower than a given threshold.

The final boundary potential is computed as:

If $B\{i,j\} > \eta(1 - \Phi_\sigma(i,j))$ then $B_{current}\{i,j\} = B_{ancient}\{i,j\} + J_i^*$.

Where $\eta$ is a parameter is a basic threshold of the edgenesse value and $\Phi_\sigma(i,j)$ is a correlation index between vessel orientations at pixels i and j defined as:

$$\Phi_\sigma(i,j)) = \frac{D_\sigma(i) \times D_\sigma(j)}{||D_\sigma(i)|| \times ||D_\sigma(j)||} \tag{9}$$

$\Phi_\sigma(i,j)$ equal to 1 if the two local direction vectors are parallel, 0 if they are orthogonal, and a value in $]0..1[$ otherwise.

Finally, min cut algorithm [13] is applied to find the segmentation with minimum energy and we keep the largest connected component in the final segmentation.

## 4    Experimental Result

We conduct experiments on the three datasets DS1, DS2 and DS3 defined in [8] with in total 91 images from 60 patients. Each dataset offers a challenging difficulty to evaluate the proposed algorithm. Experiments on Ds1 aim to evaluate the general performance of the algorithm while Ds2 contains simple vessel illumination cases and stenosis degrees. The images in DS3 present more complicated cases such as Severe stenosis, stent(s) and low signal to noise ratio. Expert hand segmentation is provided for all images of the datasets. In all experiments, $\sigma \in \{width\ of\ smallest\ vessel, ..., width\ of\ largest\ vessel\}$, in our dataset, the approximate width of vessels vary from less than 1 pixel to more than 10 pixels, $\eta = 60\%$. In Figs. 1, 2 and 3, we present original sample images respectively from DS1, DS2 and DS3 and their corresponding segmentation results with the AQCA method and the proposed method.

We can visibly notice that the proposed method can detect more vessels and is much closer to the ground truth than the AQCA method for the three datasets. In fact, the results given by the AQCA method describe the main vessels in the image and ignore others relevant ones, while our method successfully detect most of them. In order to quantitatively assess the quality of the proposed algorithm,

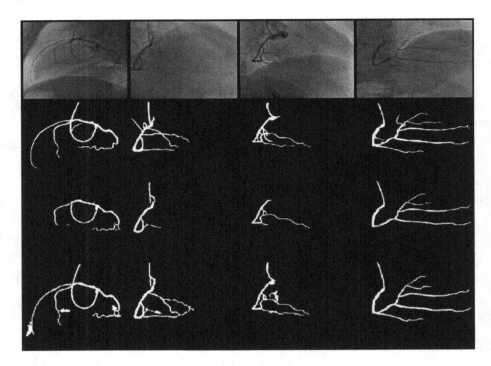

**Fig. 1.** Segmentation results: Top to bottom: real angiograms images from DS1; ground truth; AQCA and our method

we compute the false positive rate RFP, the sensibility and Dice similarity coefficient. DSC indicates the spatial cover between the segmented vessel and the ground truth and defined as:

$$DSC = (2 \times |GT \cap Seg|) / (|GT| + |Seg|) \tag{10}$$

Where $|GT|$ and $|Seg|$ are the number of pixels classified as vessels respectively in the ground truth image and segmented image. $|GT \cap Seg|$ gives the number of pixels in the intersected region between these two images.

The sensitivity is the percentage of pixels belonging to vessels which correctly segmented and defined as:

$$Sensibility = (TP) / (TP + FN) \tag{11}$$

Where $TP$ is the number of true positive cases and $FN$ is the number of false negative cases.

The false positive rate RFP is the percentage of pixels belonging to vessels which incorrectly segmented as belonging to the background and defined as:

$$RFP = (FP) / (FP + TN) \tag{12}$$

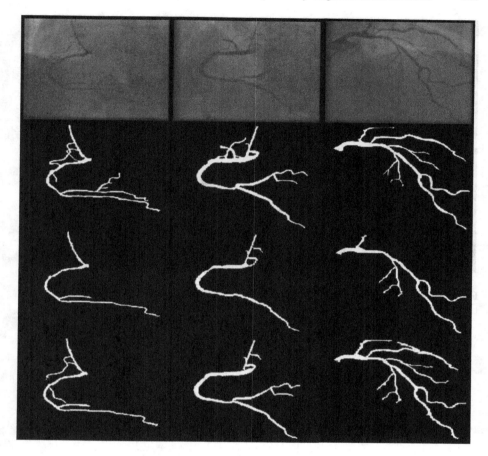

**Fig. 2.** Segmentation results: Top to bottom: real angiograms images from DS2; ground truth; AQCA and our method

where $FP$ is number of false positives and $TN$ is number of true negatives.

The obtained results are presented as boxplots that indicate the first, second and third quartile with whiskers from minimum to maximum. In Fig. 1, Boxplots of DSC measures are drawn for both AQCA and the proposed method and for each of the three datasets.

Figure 4 shows that the DSC measures given by the proposed method exceed 0.7 for Ds1 and Ds2 which is considerate as good index. The measures are less satisfying in the case of Ds3 which could be explained by the quality of the images with a very non homogeneous background. The results of the proposed method outperform the ones given by the AQCA method. This difference is more remarkable for the two first datasets but less significant for the third one (Fig. 5).

**Fig. 3.** Segmentation results: Top to bottom: real angiograms images from DS3; ground truth; AQCA and our method

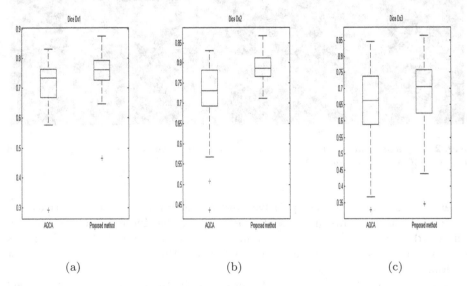

**Fig. 4.** Dice similarity measure for: (a): DS1, (b): DS2 and (c): DS3

We can clearly notice that the proposed method is more accurate than the AQCA method in fact for the three datasets, the sensitivity values of our method exceed those of the AQCA method.

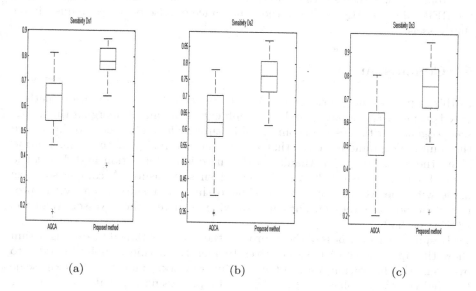

**Fig. 5.** Sensitivity measure for: (a): DS1, (b): DS2 and (c): DS3

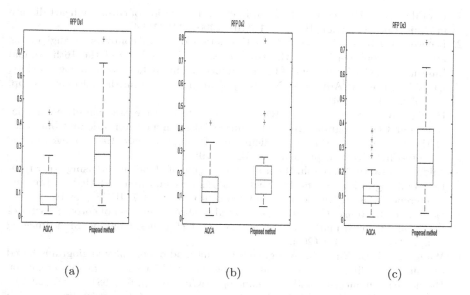

**Fig. 6.** RFP measure for: (a): DS1, (b): DS2 and (c): DS3

From Fig. 6, we can see that the RFP values of our algorithm are higher than the RFP values of AQCA due to the appearance of false positives frequently on thin vessels.

## 5  Conclusion

In this paper, we have presented a new coronary artery segmentation method based on multiscale analysis and the graph cut algorithm by merging the multiscale edgeness with Hessian geometrical features. The main idea is to introduce the multiscale boundary term that ensure the continuity of the segmentation along the artery direction thanks to the direction information and detect the vessel boundaries using the multi-scale edgeness measure. A comparison was made with the AQCA method and showed that the results given by the proposed method are very encouraging allowing to detect more vessels than the AQCA method.

Despite the robustness of the approach, the results obtained by our algorithm show the appearance of false positives frequently on thin vessels. In order to overcome this limitation, we will try, as a future work, to locally adapt the scale to avoid the overemphasis of these vessels by processing them at a scale less or equal than their sizes.

## References

1. Pyörälä, K., de Backer, G., Graham, I., et al.: Prevention of coronary heart disease in clinical practice. Eur. Heart J. **15**(10), 1300–1331 (1994)
2. Lecornu, L., Roux, C., Jacq, J.-J.: Extraction of vessel contours in angiograms by simultaneous tracking of the two edges. In: Proceedings of the 16th Annual International Conference of the IEEE Engineering in Medicine and Biology Society, 1994. Engineering Advances: New Opportunities for Biomedical Engineers, pp. 678–679. IEEE (1994)
3. Higgins, W.E., Spyra, W.J.T., Ritman, E.L.: Automatic extraction of the arterial tree from 3-D angiograms. In: Proceedings of the Annual International Conference of the IEEE Engineering in Medicine and Biology Society, 1989. Images of the Twenty-First Century, pp. 563–564. IEEE (1989)
4. Niki, N., Kawata, Y., Satoh, H., et al.: 3D imaging of blood vessels using x-ray rotational angiographic system. In: Nuclear Science Symposium and Medical Imaging Conference, 1993 IEEE Conference Record, pp. 1873–1877. IEEE (1993)
5. Bouraoui, B., Ronse, C., Baruthio, J., et al.: 3D segmentation of coronary arteries based on advanced mathematical morphology techniques. Comput. Med. Imaging Graph. **34**(5), 377–387 (2010)
6. Wang, S., Li, B., Zhou, S.: A segmentation method of coronary angiograms based on multi-scale filtering and region-growing. In: 2012 International Conference on Biomedical Engineering and Biotechnology (iCBEB), pp. 678–681. IEEE (2012)
7. Nimura, Y., Kitasaka, T., Mori, K.: Blood vessel segmentation using line-direction vector based on Hessian analysis. In: SPIE Medical Imaging. International Society for Optics and Photonics, pp. 76233Q–76233Q-9 (2010)

8. Kerkeni, A., Benabdallah, A., Manzanera, A., et al.: A coronary artery segmentation method based on multiscale analysis and region growing. Comput. Med. Imaging Graph. **48**, 49–61 (2016)
9. Kerkeni, A., Benabdallah, A., Bedoui, M.H.: Coronary artery multiscale enhancement methods: a comparative study. In: Kamel, M., Campilho, A. (eds.) ICIAR 2013. LNCS, vol. 7950, pp. 510–520. Springer, Heidelberg (2013)
10. Schmitter, D., Gaudet-Blavignac, C., Piccini, D., et al.: New parametric 3D snake for medical segmentation of structures with cylindrical topology. In: 2015 IEEE International Conference on Image Processing (ICIP), pp. 276–280. IEEE (2015)
11. Hernández-vela, A., Gatta, C., Escalera, S., et al.: Accurate coronary centerline extraction, caliber estimation, and catheter detection in angiographies. IEEE Trans. Inform. Technol. Biomed. **16**(6), 1332–1340 (2012)
12. Kolmogorov, V., Boykov, Y.: What metrics can be approximated by geo-cuts, or global optimization of length/area and flux. In: Tenth IEEE International Conference on Computer Vision (ICCV 2005), vol. 1, pp. 564–571. IEEE (2005)
13. Kolmogorov, V., Zabin, R.: What energy functions can be minimized via graph cuts? IEEE Trans. Pattern Anal. Mach. Intell. **26**(2), 147–159 (2004)
14. Liao, R., Luc, D., Sun, Y., et al.: 3-D reconstruction of the coronary artery tree from multiple views of a rotational X-ray angiography. Int. J. Cardiovasc. Imaging **26**(7), 733–749 (2010)

# Gaussian Bayes Classifier for 2D Shapes in Kendall Space

Hibat Allah Rouahi[1]([✉]), Riadh Mtibaa[1,2]([✉]), and Ezzeddine Zagrouba[1]([✉])

[1] Laboratoire LIMTIC, Institut Supérieur d'Informatique,
Université de Tunis El Manar, 2 Rue Abou Rayhane Bayrouni, 2080 Ariana, Tunisie
hiba_rouahi@yahoo.fr, rmtibaa@gmail.com,
ezzeddine.zagrouba@fsm.rnu.tn
[2] Institut Supérieur des sciences appliquées et de technologie de Sousse (ISSAT),
Université de Sousse, rue Khalifa Karoui Sahloul 4, 526, Sousse, Tunisie

**Abstract.** We propose a 2D-shape Gaussian Bayes classifier based upon Kendall's representations that help to quotient out the effects of non-altering shape geometric transformations. The Kendall space is a non linear space that coincides with the unit sphere modulo an isometry group. The proposed Riemannian metric is more apt in the case where shapes are different only in translation, scale and rotation. In addition to that, the manifold structure of this space renders the multivariate statistical analysis implementation unfeasible in practice. Consequently, tools such as learning and classification models are non trivial and not frequently available. To overcome these issues, we adapt the Gaussian Bayes classifier to this space. We computed the likelihood parameters through appropriate projections onto Kendall tangent space that provides a good linear approximation. In order to validate the robustness of our classifier, we proceeded to computer simulations using several benchmarks.

**Keywords:** Kendall space · Gaussian Bayes classifier · Likelihood parameters

## 1 Introduction

Objects'recognition is one of the most active areas of research with crucial applications in diverse fields. Object representation is a primordial step to systematically allocate mathematical structures to shapes in order to ease the implementation and analysis of the relevant classification algorithms. Objects may be described according to main parameters: colors, textures, shapes, movements and locations. However, shape remains a critical feature for object recognition as commonly accepted by the computer vision community. The need for consistent structures that help handle the rich repertoire of shapes has elicited the investigation of several researchers. Among the proposed representations are: continuous contour parametrization [1], medial [2], active model [3], and algebraic representations [4]. Once a particular representation is defined, a classifier

© Springer International Publishing AG 2017
B. Ben Amor et al. (Eds.): RFMI 2016, CCIS 684, pp. 152–160, 2017.
DOI: 10.1007/978-3-319-60654-5_13

is either trained during a learning phase through probability models and pres-elected shape sets, or designed through a clustering technique. Another type of invariant shape representation has been proposed, the Fourier descriptors along with moment invariants. These are among the main techniques, yet are only based on continuous regions and contours [5]. In 1977, D.G Kendall proposed a new invariant representation where he identified a shape with the geometri-cal information that remains after filtering out the location, scale and rotation effects from initial configuration matrices that capture objects'shapes through landmarks. The elimination of irrelevant effects generated a non-linear space. Consequently, the application of various clustering and classification methods is not appropriate. The Riemannian metric for this space refers only to whether two shapes are identical (different only in translation, rotation and scale) or not, yet in many cases we need to measure shape similarity. An appropriate metric for shape classification should not only suit certain invariance proper-ties but also satisfy the different input properties. This study tries to devise a robust classifier through the combination of an efficient shape representa-tion emerging within the framework of the Kendall's theory with the Bayesian approach renowned for its rigorous theoretical foundation and optimal classi-fication results. We based our choice of the Kendall representation upon the firm belief that its adequate implementation yields computational effectiveness. Indeed, representation-based-matrices are easy for calculate and are much less time consuming compared to those using more complex structures. Our main contribution is the adaptation of the Gaussian Bayes classifier in Kendall space and its local approximation through projections onto tangent spaces. Section 2 of this paper exposes an overview of some works related to Kendall space. The pro-posed method is presented in Sect. 3. Experiments are presented and discussed in Sect. 4. Finally, we conclude the work and suggest proposals in Sect. 5.

## 2   Previous Works on Kendall Space

The study of shapes dates back to D'Arcy Thompson, but the first more sys-tematic algorithmic treatment of shape representations and metrics is due to Bookstein and Kendall. They represented shape by a collection of ordered land-mark points invariant to Euclidean similarity transformations. Two objects have the same shape if they can be translated, scaled and rotated to each other so that they match exactly. This standpoint has led to the foundation of the Kendall shape theory [6,7] which is a most popular and widely discrete used shape repre-sentation. Number of works elaborated by shape theories experts like Bookstein [8], Dryden [9], have adopted Kendall definition to get finite dimensional spaces from landmarks coordinates. The readers are referred to [10,11] for a thorough view of the recent developments of Kendall theory. The solution to resolve the problem of non-conforming between Kendall space and the classical linear algo-rithm of classification was proposed by Jayasumana et al. in [12] and is to per-form a mapping operation from the manifold to the Hilbert space using a kernel function. Which produces a richer representation of the data, and makes tasks such as classification easier. However, only positive definite kernels yield a mapping to a Hilbert space and a poor choice of kernel can often result in reduced classification performance. Another idea to adapt an unsupervised learning algorithm, k-means, to Kendall space is to integrate the Procrustes mean [13]. Here we focus on adapting the Bayesian approach to Kendall space through a linear tangent space.

# 3 Our Approach of Supervised Learning in Kendall Space

## 3.1 Landmarks Selection

At the outset we fixed the value of the number $N_c$ of the targeted shape classes $\omega_i$, for $1 \leq i \leq N_c$, since we are working within a supervised context. Then, we generated for each $\omega_i$ amongst the $N_c$ classes a set $\{X_j^{i\star}\}_{1 \leq j \leq N_{ps}}$ comprising $N_{ps}$ initial configurations of learning samples. Here, we gathered in each matrix $X_j^{i\star}$ the vector column coordinates of some fixed number of labelled landmarks, which are selected to capture the shape of the jth sample from the class $\omega_i$. More accurately, the selection consists of picking up $k$ landmarks from the contours. We noticed that some of the benchmarks also provide a set of landmarks selected by experts, unless said landmarks were not valid, in that case we performed the selection task manually based on templates assigned to each class.

## 3.2 Learning Shapes Processing

Let $X_j^{i\star} = \begin{pmatrix} x_0^\star & x_1^\star & \dots & x_{k-1}^\star \end{pmatrix}$ be one of the learning samples in $\{X_j^{i\star}\}_{1 \leq j \leq N_{ps}}$ for the class $\omega_i$ (Cf. Sect. 3.1). We started by calculating the center of $X_j^{i\star}$ to remove the translation effect from $X_j^{i\star}$, by moving $x_c^\star$ to the origin of the coordinates system, when each $x_\ell^\star$ is substituted with $x_\ell^\star - x_c^\star$, for all $0 \leq \ell \leq k-1$, respectively.

$$x_c^\star = \frac{1}{k} \sum_{\ell=0}^{k-1} x_\ell^\star \tag{1}$$

Next, we performed a dimension reduction via a right multiplication of the last version of $X_j^{i\star}$ by the recentering orthogonal matrix $Q$,

$$Q = \begin{cases} Q_{\ell 1} = \frac{1}{\sqrt{k}}, & 1 \leq \ell \leq k; \\ Q_{\ell\ell} = \frac{\ell-1}{\sqrt{\ell(\ell-1)}}, & 2 \leq \ell \leq k; \\ Q_{\ell h} = -\frac{1}{\sqrt{\ell(\ell-1)}}, & 1 \leq \ell \leq h-1, 2 \leq h \leq k; \\ Q_{\ell h} = 0, & \text{otherwise.} \end{cases} \tag{2}$$

to get an intermediate representation matrix $\tilde{X}_j^i$ with general form $\begin{pmatrix} 0 & \tilde{x}_1 & \tilde{x}_2 & \dots & \tilde{x}_{k-1} \end{pmatrix}$ which we reduced naturally to

$$\tilde{X}_j^i = \begin{pmatrix} \tilde{x}_1 & \tilde{x}_2 & \dots & \tilde{x}_{k-1} \end{pmatrix} \tag{3}$$

After that, we eliminated the scaling effect through normalization to get

$$X_j^i = \frac{1}{\sqrt{tr(\tilde{X}_j^i(\tilde{X}_j^i)^t)}} \tag{4}$$

The set $\{X_j^i\}_{1 \leq j \leq N_{ps}}$ coincides with points on the $2(k-1)-1$ dimensional unit sphere, denoted by $\mathcal{S}_2^k$. We notice that any matrix $TX_j^i$, for $T \in \mathbf{SO}(2)$, has the same shape as $X_j^i$. So, from now on, $X_j^i$ is treated as a pre-shape representation and the sought shape denoted by $\pi(X_j^i)$ identifies an equivalence class modulo the left action of rotations $T$ in $\mathbf{SO}(2)$ on the pre-shape $X_j^i$; the space of all possible shapes is the $2(k-1)-2$ dimensional Kendall space, which is the quotient space.

$$\Sigma_2^k = \mathcal{S}_2^k / \mathbf{SO}(2) \tag{5}$$

The pseudo-singular values decomposition helps to write any pre-shape $X_j^i$ as a three-factors product,

$$X_j^i = U(\Lambda\, 0)V \tag{6}$$

where $U \in \mathbf{SO}(2)$, $V \in \mathbf{SO}(k-1)$, $0$ is the null matrix of dimensions $2 \times (k-3)$, and $\Lambda$ is the $2 \times 2$ diagonal matrix $diag\{\lambda_1, \lambda_2\}$ such that $\lambda_1 \geq |\lambda_2|$, $\lambda_1^2 + \lambda_2^2 = 1$, and $\lambda_2 \geq 0$ unless $k = 3$. This decomposition provides a systematic way to decide whether or not any couple of learning pre-shapes belong to the same equivalence class. In order to quotient out the left acting orthogonal matrices to obtain the learning shapes in $\Sigma_2^k$ (5), we calculated $U_{X_{j_1}^i}(\Lambda_{X_{j_1}^i}\, 0)V_{X_{j_1}^i}$ and $U_{X_{j_2}^i}(\Lambda_{X_{j_2}^i}\, 0)V_{X_{j_2}^i}$ of each couple of pre-shapes $X_{j_1}^i$ and $X_{j_2}^i$ in $\{X_j^i\}_{1 \leq j \leq N_{ps}}$ (6). Then, we decided that $\pi(X_{j_1}^i)$ and $\pi(X_{j_2}^i)$ are identical if and only if $\Lambda_{X_{j_1}^i} = \Lambda_{X_{j_2}^i}$ and both first rows of $V_{X_{j_1}^i}$ and $V_{X_{j_2}^i}$ are exactly the same, the remaining $k-3$ rows of $V_{X_{j_1}^i}$ and $V_{X_{j_2}^i}$ do not matter since they are multiplied by the null matrix of dimensions $2 \times (k-3)$ appearing in $(\Lambda_{X_{j_1}^i}\, 0)$ and $(\Lambda_{X_{j_2}^i}\, 0)$, respectively. Naturally, we did not care about the left acting orthogonal matrices $U_{X_{j_1}^i}$ and $U_{X_{j_2}^i}$ because they do not affect shapes. This way, we succeeded in regrouping the $N_{ps}$ learning pre-shapes of $\{X_j^i\}_{1 \leq j \leq N_{ps}}$ into $N_s^i$ learning equivalence classes. Concretely, for all $1 \leq p \leq N_s^i$, the pth equivalence class is represented by any candidate denoted by $\pi(X_p^i)$, and qualified as the learning shape.

## 3.3   Likelihood for Parameters Inference

We describe here, our approach of learning procedure to estimate the expectation vector $\mu_i$ as well as the covariance matrix $\Sigma_i$ of the Gaussian likelihood. We assume that the learning samples in each set $\{\pi(X_p^i)\}_{1 \leq p \leq N_s^i}$ are independent and identically distributed according to the likelihood of each $\omega_i$, respectively. In a first step, we calculated the mean shape $\pi(\hat{\nu}_i)$ of the set of learning shapes

$\left\{\pi\left(X_p^i\right)\right\}_{1\leq p\leq N_s^i}$ which we used later as a reference shape where we approximated $\Sigma_2^k$ locally by its tangent space [14]. Specifically, we looked for $\pi\left(\hat{\nu}_i\right)$ as a solution of the minimization problem

$$\arg\inf_{\pi(\nu)\in\Sigma_2^k}\sum_{p=1}^{N_s^i} d_F^2\left(\pi\left(X_p^i\right),\pi\left(\nu\right)\right) \tag{7}$$

which involves the procrustes distance between $\pi\left(X_p^i\right)$ and $\pi\left(\nu\right)$

$$d_F^2\left(\pi\left(X_p^i\right),\pi\left(\nu\right)\right) = \sin\left(\rho\left(\pi\left(X_p^i\right),\pi\left(\nu\right)\right)\right) \tag{8}$$

Here, $\rho$ is the distance function defined for $\Sigma_2^k$ as

$$\rho\left(\pi\left(X_p^i\right),\pi\left(\nu\right)\right) = \arccos\left(\lambda_1+\lambda_2\right) \tag{9}$$

with the pseudo-singular values $\lambda_1$ and $\lambda_2$ of $X_p^i\nu^t$ for arbitrary pre-shapes $X_p^i$ and $\nu$ of $\pi\left(X_p^i\right)$ and $\pi\left(\nu\right)$, respectively. In a second step, we mapped each learning shape $\pi\left(X_p^i\right)$ in $\left\{\pi\left(X_p^i\right)\right\}_{1\leq p\leq N_s^i}$ onto its projection $\bar{\pi}_i\left(X_p^i\right)$ computed as

$$\bar{\pi}_i\left(X_p^i\right) = \left(I_m - \pi\left(\hat{\nu}_i\right)\pi\left(\hat{\nu}_i\right)^t\right)\pi\left(X_p^i\right), \tag{10}$$

In practice, we computed $\pi\left(\hat{\nu}_i\right)$ and $\bar{\pi}_i\left(X_p^i\right)$ for all $1\leq i\leq N_c$ and $1\leq p\leq N_s^i$ using the generalised procrustes analysis [15]. We emphasize here on two important features of the last function: first, we get always the same mean shape $\pi\left(\hat{\nu}_i\right)$ even if we use left rotated versions of the actual shapes $\pi\left(X_p^i\right)$, for all $1\leq i\leq N_c$, respectively. Second, all the left rotated versions of the shape $\pi\left(X_p^i\right)$ project always to the same point of the tangent space to $\Sigma_2^k$ at $\pi\left(\hat{\nu}_i\right)$. In a last step, we reshaped each one of the matrices $\bar{\pi}_i\left(X_p^i\right)$ onto a row vector $\bar{\pi}_i^v\left(X_p^i\right)$. Then, we profited from the maximum likelihood method to get likelihood parameters

$$\mu_i = \frac{1}{N_s^i}\sum_{p=1}^{N_s^i}\bar{\pi}_i^v\left(X_p^i\right), \tag{11}$$

$$\Sigma_i = \frac{1}{N_s^i}\sum_{p=1}^{N_s^i}\left(\bar{\pi}_i^v\left(X_p^i\right)-\mu_i\right)^t\left(\bar{\pi}_i^v\left(X_p^i\right)-\mu_i\right), \tag{12}$$

Both parameters dimensions are $1\times 2\left(k-1\right)$ and $2\left(k-1\right)\times 2\left(k-1\right)$, respectively.

## 3.4   Generalization

In the present subsection we detail how we conducted the classification task of general shapes. After obtaining the likelihood phase (Cf. Sect. 3.3), we used the values of the Gaussian likelihood $\mu_i$ and $\Sigma_i$ of each class $\omega_i$ to compute the posteriori probabilities

$$p\left(\omega_i|\pi\left(X\right)\right) = \frac{p\left(\bar{\pi}_i^v\left(X\right)|\omega_i\right)P\left(\omega_i\right)}{p\left(\bar{\pi}_i^v\left(X\right)\right)} \tag{13}$$

where

$$p\left(\bar{\pi}_i^v\left(X\right)|\omega_i\right) = \frac{1}{\left(2\pi\right)^{(k-1)}\sqrt{\det\left(\Sigma_i\right)}}\exp\left(-\frac{1}{2}\left(\bar{\pi}_i^v\left(X\right)-\mu_i\right)\Sigma_i^{-1}\left(\bar{\pi}_i^v\left(X\right)-\mu_i\right)^t\right), \tag{14}$$

$P\left(\omega_i\right)$ is an a priori probability of $\omega_i$, and $p\left(\bar{\pi}_i^v\left(X\right)\right)$ is the evidence term. Finally, the maximum amongst the values of $p\left(\omega_i|\pi\left(X\right)\right)$ indicates the class of $\pi\left(X\right)$. Here again, we draw the attention of the readers that if we represent the shape $\pi\left(X\right)$ by any $TX$ where $T \in \mathbf{SO}\left(2\right)$, then we still have exactly the same likelihood value in (14), because all such $TX$ matrices project always through (10) to the same $\bar{\pi}_i\left(X\right)$ in the tangent space to $\Sigma_2^k$ at $\pi\left(\hat{\nu}_i\right)$; that is to remind that our a posteriori probability distribution does depend on shapes rather than pre-shapes.

## 4    Experiments and Results Analysis

We ran several experiments for the purpose of evaluating the behavior of our classifier on $\Sigma_2^k$ (see Sect. 3 above). We used mainly four 2D shape benchmarks namely, MPEG-7, Swedish leaves [14], great apes data, and T2 mouse vertebrae data [15]. Our classifier faced an important challenge since the handled shapes correspond to domestic objects, leaves, skulls or even mouse vertebrae. The MPEG-7 offers 70 classes and 20 images per class, yet in our experiments we only used 60 classes. Figure 1 contains some annotated samples from MPEG-7. We assigned 10 images per class to the learning phase, and the rest for generalization. We will now demonstrate the benefits of our Classifier on the problem of leaf identification. We employed the Swedish leaves dataset, which contains 15 different classes with 75 images per class. We used 35 images for training and the rest for testing. Some annotated samples are shown in Fig. 2. The apes database contains the skulls of 167 specimens of great apes, with different species and both sexes: chimpanzee (26 females and 28 males), gorilla (30 females and 29 males) and orangutan (30 females and 30 males). We used 15 samples of each class for the learning phase, the rest were used for the test. Finally, the T2 mouse vertebrae database contains the second thoracic vertebra of three groups of mice: large (23 samples), small (23 samples) and control group (30 samples). We used 13 samples for each group for the learning stage and the reset for generalization.

In Table 1 we summarize the outcomes of the experiments on the aforementioned benchmarks, where the numbers of landmarks are 12 and 27 for MPEG-7 and Swedish leaves respectively, which we selected since they are not available as shown in the table. For the case of apes and T2 mouse vertebrae, the landmarks

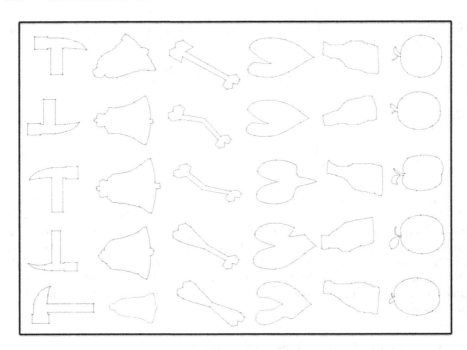

**Fig. 1.** Samples of hammer, bell, bone, heart, bottle, and apple shapes from MPEG-7. The red dots indicate the selected landmarks for each shape sample. (Color figure online)

have been provided by an expert. We end the current subsection by comparing the results of our classifier in $\Sigma_2^k$ to those of the Gaussian Bayes classifier in Euclidean space and to the kernel SVM classifiers in Hilbert space. It should be noted that the result of the last classifier is derived from this paper [12]. We gathered the results of generalization of all these classifiers in Table 2 where it is clear that our classifier outperforms the classical one, proof of its robustness.

We expect that the improvement in classification results comes from the nonexistence of redundancy within the learning sets $\left\{\pi\left(X_p^i\right)\right\}_{1\leq p\leq N_s^i}$ used by our classifier in Kendall space, compared to the sets $\left\{X_j^{i\star}\right\}_{1\leq j\leq N_{ps}}$ used by the classical classifier. Here, the nonexistence of the redundancy stems straightforwardly from the systematic elimination of translation, scale, and rotation effects during the construction of learning shapes. Besides, the complexity of our classifier in Kendall space amounts to an order determined mainly by the $2\left(k-1\right)\sum_{i=1}^{N_c}N_s^i$ coordinates values involved in learning phase, which should be much smaller than the $2kN_cN_{ps}$ coordinates values involved in the learning phase of the classical classifier since $N_s^i \ll N_{ps}$, for all $1 \leq i \leq N_c$, when the initial learning data sets get larger.

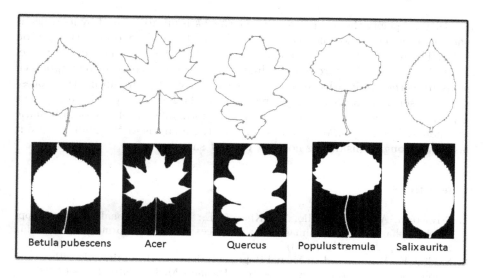

**Fig. 2.** Swedish leaves dataset samples.

**Table 1.** Computer simulations summary

| Benchmark | Landmarks | Availability | Success rate |
|---|---|---|---|
| MPEG-7 | 12 | No | 97.16% |
| Swedish leaves | 30 | No | 91.16% |
| Apes | 8 | Yes | 98.70% |
| T2 vertebrae | 6 | Yes | 97.29% |

**Table 2.** Comparison between our classifier and two other supervised classifiers

| Benchmark | Gaus-Bay in Ken | Gaus-Bay in Eucl | SVM in Hilb |
|---|---|---|---|
| MPEG-7 | 97.16% | 86.33% | 96.57% |
| Swedish leaves | 91.16% | 90.37% | 91.47% |
| Apes | 98.70% | 90.90% | — |
| T2 vertebrae | 97.29% | 78.37% | — |

## 5    Conclusion and Future Works

In the present paper we proposed an approach of supervised learning for shape classification in $\Sigma_2^k$. The supervised learning has concerned the inference of the expectation vectors and covariance matrices of the Gaussian distributions. The inference here has been realized in the context of maximum likelihood, using the available learning shapes. We detailed the procedure that helped us to construct a learning shape set from initial configurations. Then, we used the Riemannian structure to specify the mean shape where the $\Sigma_2^k$ is approximated by its tangent space. Consequently, we succeeded in establishing a likelihood density function on $\Sigma_2^k$. The results of the experiments has confirmed the robustness of our model where the success rates are outstanding. The major drawback of our approach comes from the landmarks selection which is at best a semi-automatic process where the selection remains subjective because the human supervision is necessary. In our future work, we propose to use a deformable templates for automatic detection of landmarks. We will also further the analysis of the abilities of our classifier through considering noisy or corrupt samples.

# References

1. Srivastava, A., Liu, X., Mio, W., Klassen, E.: A computational geometric approach to shape analysis in images. In: Advances in Neural Information Processing Systems, pp. 1579–1586 (2003)
2. Siddiqi, K., Pizer, S. (eds.): Medial Representations: Mathematics, Algorithms and Applications, vol. 37. Springer Science & Business Media, Netherlands (2008)
3. Cootes, T.F., Taylor, C.J., Cooper, D.H., Graham, J.: Active shape models-their training and application. Comput. Vis. Image Underst. **61**(1), 38–59 (1995)
4. Subrahmonia, J., Cooper, D.B., Keren, D.: Practical reliable Bayesian recognition of 2D and 3D objects using implicit polynomials and algebraic invariants. IEEE Trans. Pattern Anal. Mach. Intell. **18**(5), 505–519 (1996)
5. Zhang, J., Zhang, X., Krim, H., Walter, G.G.: Object representation and recognition in shape spaces. Pattern Recogn. **36**(5), 1143–1154 (2003)
6. Kendall, D.G., Barden, D., Carne, T.K., Le, H.: Shape and Shape Theory, vol. 500. Wiley, Chichester (1999)
7. Kendall, D.G.: Shape manifolds, procrustean metrics and complex projective spaces. Bull. Lond. Math. Soc. **16**(2), 81–121 (1984)
8. Bookstein, F.L.: Landmark methods for forms without landmarks: morphometrics of group differences in outline shape. Med. Image Anal. **1**(3), 225–243 (1997)
9. Dryden, I.L., Mardia, K.V.: Statistical Shape Analysis, vol. 4. Wiley, Chichester (1998)
10. Amor, B.B., Su, J., Srivastava, A.: Action recognition using rate-invariant analysis of skeletal shape trajectories. IEEE Trans. Pattern Anal. Mach. Intell. **38**(1), 1–13 (2016)
11. Giebel, S.M., Schiltz, J., Schenk, J.P.: Statistical shape analysis for the classification of renal tumors affecting children. Pak. J. Statist. **29**(1), 129–138 (2013)
12. Jayasumana, S., Salzmann, M., Li, H., Harandi, M.: A framework for shape analysis via Hilbert space embedding. In: Proceedings of the IEEE International Conference on Computer Vision, pp. 1249–1256 (2013)
13. Vinu, G., Sim, A., Alemany, S.: The k-means algorithm for 3D shapes with an application to apparel design. Adv. Data Anal. Classif. **10**(1), 103–132 (2014)
14. Soderkvist, O.: Computer vision classification of leaves from swedish trees. Masters thesis, Linkoping University (2001)
15. Dryden, I.L.: Shapes package. R Foundation for Statistical Computing, Vienna, Austria, Contributed package (2015). https://www.maths.nottingham.ac.uk/personal/ild/shapes

# Multimodal Image Fusion
# Based on Non-subsampled Shearlet
# Transform and Neuro-Fuzzy

Haithem Hermessi[✉], Olfa Mourali, and Ezzeddine Zagrouba

Research Team SIIVA - LIMTIC Laboratory,
Higher Institute of Computer Science,
University of Tunis El Manar, Tunis, Tunisia
hermessi.haithem@gmail.com, olfa.mourali@yahoo.fr,
e.zagrouba@gmail.com

**Abstract.** Due to the appealing advantages in term of medical decision making, the problem of multimodal medical image fusion has received focused research over the recent years. Moreover, complimentary imaging modalities such as CT and MRI are able to improve medical reliability by reducing uncertainty. In this paper, we propose a new algorithm for multimodal medical image fusion based on non-subsampled shearlet transform (NSST) and neuro-fuzzy. Firstly, CT and MR source images are decomposed using the NSST to obtain low and high frequency sub-bands. Maximization of absolute value is performed to fuse low frequency coefficients while high frequency coefficients are fused using the neuro-fuzzy approach. Finally, the inverse NSST is performed to gain the fused image. To assess the performance of the proposed method, several experiments are carried on different medical CT and MR image datasets. Subjective and objective assessments reveal that the proposed scheme produces better results in various quantitative criterions compared to other existing methods.

**Keywords:** Multimodal image fusion · Non-subsampled shearlet transform · Neuro-fuzzy

## 1 Introduction

Image fusion is the procedure consisting of registering and combining two or more source images to obtain single image by using image processing techniques. Its main goal is to provide suitable information for human visual perception and to reduce redundancy [1] by storing a single fused image instead of multiple source images. Image fusion technology as one of the major research fields in image processing has been applied in large scale of applications such as remote sensing, computer vision and medical diagnosis. Due to the advent of disease, complementary information is required from different imaging modalities such as magnetic resonance images (MRI), computed tomography (CT), positron emission tomography (PET) and ultrasound (US) and which the selection depends on clinical requirements like the organ undergo study. Thus, multimodal medical image fusion techniques have shown notable achievement in improving accuracy of decisions in the field of medical diagnosis and treatment planning.

© Springer International Publishing AG 2017
B. Ben Amor et al. (Eds.): RFMI 2016, CCIS 684, pp. 161–175, 2017.
DOI: 10.1007/978-3-319-60654-5_14

Image decomposition is an important tool that affects the fusion quality. Recently, multi-scale decomposition based image fusion methods has been widely used in the medical image fusion area, and has achieved great success. Wavelet theory has emerged since the beginning of the last century as a signal processing tool then directed towards image processing [2]. It has been applied for multimodal medical image fusion [3] and accomplished favorable outcome since it preserves different frequency information and allows localization both in time and spatial frequency domain. Owing to the limitation of capturing directional information, wavelets are not optimally efficient in representing images containing sharp transitions such as edges. In the past few years, multi scale geometric analysis (MGA) methods have been reported in the literature as revolutionary algorithms to overcome this deficiency. Many MGA tools have been introduced into medical image fusion for the purpose such as contourlet, ridgelet, bandelet, curvelet, etc. Those approaches have proved directional sensitivity and efficiency when dealing with medical imaging fusion process based on contourlet transform [4], non-subsampled contourlet transform [5], ridgelet transform [6], bandelet transform [7], curvelet transform [8]. Literature has reported that curvelets is an efficient transform to represent images with smooth edges similarly to contourlets which is purely discrete filter bank variety of curvelet [10]. However, multi resolution representation of the geometry cannot be provided by curvelet transform which cannot be built in the discrete domain. Moreover, contourlet transform suffer from the lack of shift-invariance [16] which was settled by non-subsampled contourlet transform but still suffering from limited number of directions and high computational cost. In recent past, shearlet theory as an extension of the wavelet framework has been provided by Labate et al. [9, 11]. It owns the advantageous properties of all above approaches and additionally it is equipped by rich mathematical structure suitable for multi resolution analysis which is very useful in for developing fast algorithmic implementations. The fact that there is no limitation on the number of directions obtained by applying the shear matrix makes the shearlet advantageous over the contourlet. Thus, shearlets build a tight frame at different scales and directions convenient to optimal sparse representation of images with edges [10]. On the other hand, Easley et al. [11] introduced the non-subsampled shearlet transform (NSST) to fill the need of shift invariance property.

Although the shearlet transform provides an efficient tool for image decomposition, one open problem that remains under investigation is how to select the appropriate fusion rules for low frequency and high frequency coefficients. The computational intelligent systems play a crucial role in the field of medicine. In [27] a method based on fuzzy classification and regions segmentation is proposed to detect tumoral zone in the brain IRM images. Besides, Neuro-fuzzy logic is one of the approaches which are finding applications in image processing fields as well as in medical image fusion [26]. As a fusion rule, it consists of a combination of artificial neural network (ANN) and fuzzy logic where neurons can be trained and the membership functions can also be applied for decision making. Neuro-fuzzy inference system (NFIS) has been adopted in [17] to fuse multimodal medical images. The recent literature in [18, 19] have also reported the combination of multi scale geometric analysis with neuro-fuzzy logic in the purpose to fuse medical images. In [18], Das et al. employed the non-subsampled contourlet transform to decompose input images and reduced pulse coupled neural network with fuzzy logic is utilized as a fusion rule. Furthermore, in [19] images are decomposed using wavelet transform then fused based on neuro-fuzzy. In this work, MRI-CT image fusion is performed in order to help as an accurate tool for planning the correct surgical procedure or therapy. In this regard, we firstly propose to decompose input CT and MRI images using the shearlet transform. Then, we perform neuro-fuzzy inference to fuse high sub band similarly to low sub band given by the shearlet decomposition.

The remainder of this paper is organized as follows: recent literature associated to shearlet transform and neuro-fuzzy in the realm of medical image fusion is described in Sect. 2. In Sect. 3, we present the proposed fusion method consisting of shearlet decomposition of the input images fused based on NFIS. Experimental results and comparisons are discussed in the last section. Finally, conclusion and future research directions are outlined.

## 2 Related Works

Shearlet transform (ST) is equipped with rich mathematical structure which is improved by shearing filters having small support size then directional filters so it can be implemented more efficiently. Shearlet theory has been studied and applied gradually. Its applications in image processing were extended to image denoising [11] and edge detection [12] where it has been shown that ST allows one to exactly identify the location and the orientation of the edge. However, this broader area of research at the cross road of medical image fusion is still under exploring. ST was introduced by Miao *et al.* [13] in the field of image fusion and accomplished satisfying performance. Deng *et al.* [14] also applied ST to fuse remote sensing images but still not able to overcome the problem of shift invariance. Another extension provided by Wang *et al.* [15], the sift-invariant sheralet transform (SIST) which is combined with Hidden Markov Tree (HMT) to model the dependent relationship for the SIST sub-bands. Owning the property of shift invariance the non-subsampled shearlet transform combined with neural networks was conducted by Kong and Liu in [23]. In [16], a fusion method for the CT and MRI images were presented utilizing pulse coupled neural network in the non-subsampled shearlet transform (NSST) domain which incorporate several different combination of the shearing with the non-subsampled Laplacian pyramid transform. It has been concluded that NSST can suppress the pseudo-Gibbs phenomenon advantageously over standard shearlet. Furthermore, MGA tools have been proposed in junction with neuro-fuzzy [18, 19]. Moreover, non-subsampled contourlet transform are applied to decompose input images into low and high frequency sub-bands, and then the neuro-fuzzy is performed as a fusion rule [25]. In [20], Rajkumal *et al.* compared lifting wavelet transform and neuro-fuzzy with only iterative neuro-fuzzy and concluded the superiority of the second approach.

## 3 Proposed Fusion Methods

The task of multimodality image fusion is to make many salient features in the new image such as regions and their boundaries. However, image registration is an important requirement applied for fusion technique. In this paper, it is assumed that the source images are registered before initiating the fusion process. In the following, we propose to decompose the CT and MRI images using the shearlet transform to obtain low and high frequency coefficient. Then, low frequency coefficients are fused by maximization of absolute value while high frequency sub band is fused based on NFIS.

### 3.1 Non Subsampled Shearlet Transform

In contrast to all MGA tools, the shearlet provides a unique combination of mathematical rigidness and computational efficiency when addressing edges. Proposed by K. Guo *et al.* [9, 11, 12, 21, 22], it is derived from the theory of wavelets. In dimension $n = 2$, the affine systems with composite dilation are defined as follows:

$$A_{AS}(\Psi) = \left\{ \Psi_{j,l,k}(x) = |\det A|^{j/2} \Psi(S^l A^j x - k); j, l \in \mathbb{Z}, k \in \mathbb{Z}^2 \right\} \tag{1}$$

Where $\Psi \in L^2(\mathbb{R}^2)$, A, S are both $2 \times 2$ invertible matrices, and $\det|S| = 1$. The elements of this system are called composite wavelet if $A_{AS}(\Psi)$ forms a tight frame for $L^2(\mathbb{R}^2)$ satisfied by:

$$\sum_{j,l,k} |\langle f, \Psi_{j,l,k} \rangle|^2 = \|f^2\| \tag{2}$$

The shearlet transform is a function of three variables: the scale $j$, the shear $l$ and the translation $k$. Let $A$ denote the scaling matrix and $S$ stand for the shear matrix. For each $a > 0$ and $s \in \mathbb{R}$,

$$A = \begin{pmatrix} a & 0 \\ 0 & \sqrt{a} \end{pmatrix}, \quad S = \begin{pmatrix} 1 & s \\ 0 & 1 \end{pmatrix} \tag{3}$$

The matrices described above plays an important role in the process of the shearlet transform. The former matrix A controls the scale of the shearlet by applying a fine dilation along the two axes which increasingly elongated the frequency support at fine scales. The latter matrix, which is not expensive, dominates the orientation of the shearlet. The tiling of the frequency and the size of frequency support are illustrated in Fig. 1 for a particular values of $a$ and $s$. The frequency support size of the shearlet for particular values of a and s is shown in Fig. 2.

In reference [11], commonly assume $a = 4, s = 1$, where $A_0$ and $S_0$ are respectively the anisotropic dilation matrix and the shear matrix. Equation (3) gives:

$$A_0 = \begin{pmatrix} 4 & 0 \\ 0 & 2 \end{pmatrix}, \quad S_0 = \begin{pmatrix} 1 & 1 \\ 0 & 1 \end{pmatrix}$$

For any $\xi = (\xi_1, \xi_2) \in \widehat{\mathbb{R}}^2, \xi_1 \neq 0$, let $\psi^{(0)}(\xi)$ be given by

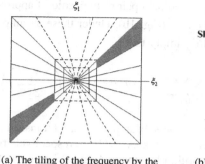

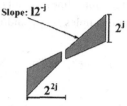

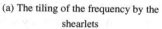

(a) The tiling of the frequency by the shearlets

(b) The size of the frequency support of a shearlet

**Fig. 1.** The structure of frequency tiling and the size of the frequency support.

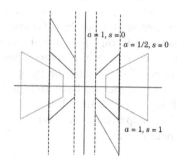

**Fig. 2.** Frequency support of shearlets $\psi_{j,l,k}$ for different values of $a$ and $s$.

$$\hat{\psi}^{(0)}(\xi) = \hat{\psi}^{(0)}(\xi_1, \xi_2) = \hat{\psi}_1(\xi_1)\hat{\psi}_2(\xi_2/\xi_1)$$

where $\hat{\psi}_1, \hat{\psi}_2 \in \mathbf{C}^\infty(\widehat{\mathbb{R}})$ are both wavelets, and supp $\hat{\psi}_1 \subset [-1/2, -1/16] \cup [1/16, 1/2]$, supp $\hat{\psi}_2 \subset [-1, 1]$. In addition, assume that:

$$\sum_{j\geq 0} \left|\hat{\psi}_1\left(2^{-2j}\omega\right)\right|^2 = 1 \text{ for } |\omega| \geq \frac{1}{8} \tag{4}$$

and, for each $j \geq 0$,

$$\sum_{l=-2^j}^{2^j-1} \left|\hat{\psi}_2\left(2^j\omega - l\right)\right|^2 = 1, \quad |\omega| \leq 1 \tag{5}$$

That is, each element $\psi_{j,l,k}$ is supported on a pair of trapezoid of approximate scope $2^{2j} \times 2^{j}$ oriented along lines of slope $l2^{-j}$ (Fig. 1b). Under these assumptions (Eqs. 4 and 5), several examples of $\hat{\psi}_1$ and $\hat{\psi}_2$ imply that:

$$\sum_{j \geq 0} \sum_{l=-2^j}^{2^j-1} \left| \hat{\psi}^{(0)} \left( \xi A_0^{-j} B_0^{-l} \right) \right|^2 = \sum_{j \geq 0} \sum_{l=-2^j}^{2^j-1} \left| \hat{\psi}_1 \left( 2^{-2j} \xi_1 \right) \right|^2 \left| \hat{\psi}_2 \left( 2^j \frac{\xi_1}{\xi_2} - l \right) \right|^2 = 1 \quad (6)$$

Accordingly, we can obtain discrete non-subsampled shearlet transform by sampling the shearlet on a proper discrete set. Suppose that A and B are respectively two registered CT and MRI images. Our fusion algorithm for image A and B begins with performing discrete NSST for these two images to obtain low-and-high frequency sub band coefficients of them as illustrated in Fig. 3. The image decomposition process is divided in two steps: non-subsampled pyramid (NSP) is used to accomplish multi-scale factorization by applying non-subsampled filter banks in order to satisfy shift-invariance. The decomposition leads to $j + 1$ sub images; one is low frequency image, the others represent the high frequency images where j denotes the number of decomposition levels. The second phase performs the multi-directional decomposition realized by the shearing filters (SF) at each scale which induces directional details information. Orientation factorization with l stages in high frequency produces $2^l$ directional sub images. The NSST decomposition process is illustrated in Fig. 4 where the two basic steps are demarcated. In this work, decomposition level by NSP is $j = 3$ and the sub-band filter adopted is "*maxflat*" in a purpose to be aligned with the compared methods based on NSST [16, 26, 28] and to investigate the efficiency of neuro-fuzzy.

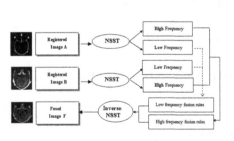

**Fig. 3.** Block diagram of the proposed fusion method

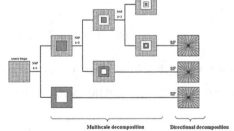

**Fig. 4.** Schematic diagram of multi-scale and multidirectional decomposition of NSST

### 3.2 Neuro-Fuzzy Inference System Based Image Fusion

Properties like brightness and edges have fuzzy effects on images due to the non-uniform illumination and inherent image vagueness [24]. NFIS is a feed-forward neural network system in which neural nets are used to tune the membership functions of fuzzy sets that operate as a decision making system [19]. The main concept of fuzzy logic lies in fact that fuzzy sets are defined by a membership function which associate a

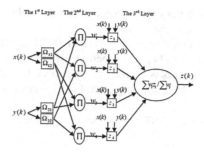

**Fig. 5.** The schematic framework of NFIS

membership degree for each element of those sets. The hybrid technique is performed in three steps; first, the membership function and fuzzy rules are defined and adaptive neuor-fuzzy inference system (ANFIS) is generated from the training data using "genfis1" function which provides initial conditions for the training. The second step consists of routine training for Sugeno-type fuzzy inference system using "anfis" function in a regard to identify the membership parameters. Finally, the total output is calculated. The process learning of NFIS and its structure are illustrated in Fig. 5. Three layers are involved; the first calculates the input membership degree, the second, calculates the pertinence degree of each rule and the final adds up the output of NFIS.

**Low and high frequency fusion rules.** Low frequency coefficients of the fused image are conventionally given by the averaging method. However, this technique is only able to contribute with low contrast result [4]. To overcome this deficiency, low frequency sub-bands of input images are chosen to be fused using the maximum of the absolute value to preserve more contrast. Thus, $LF_{A,B}(i,j)$ denotes the low frequency coefficients located at $(i,j)$ of image A or B, the fused low sub-band is given as follows:

$$LF_F(i,j) = \begin{cases} LF_A(i,j)|LF_A(i,j)| \geq |LF_B(i,j)| \\ LF_B(i,j)|LF_A(i,j)| < |LF_B(i,j)| \end{cases} \tag{6}$$

On the other hand, high frequency coefficients are fused based on neuro-fuzzy approach. At each decomposition level and for each sub-image obtained by the shearing filter, NFIS is performed in a goal to fuse the trained inputs. The fusion process is described in the following.

**Algorithm.** The proposed medical image fusion method as illustrated in Fig. 3 pursue the following steps:

- *Step 1*: Pre-registered CT and MR image are decomposed by NSST to obtain low and high frequency coefficients.
- *Step 2*: Low frequency coefficients of the source images are fused by the greatest of the absolute value methods (Eq. 6).
- *Step 3*: High frequency coefficients of each decomposition level and each sub-image are fused based on NFIS as follows:

- *Step 3.1*: Form a training data in column and reshaping the input sub-images in column form to get the check data.
- *Step 3.2*: Generate fuzzy inference system (FIS) structure from training data, number and type of membership function using "genfis1" command.
- *Step 3.3*: Training process is performed by applying "anfis" command involving the generated FIS and the training data. Finally, the fuzzy inference calculation is performed.
- *Step 4*: Apply the inverse NSST on the fused coefficients to get the fused medical image.

## 4 Experimental Results and Comparisons

In this section, several illustrative experiments are conducted in order to assess the effectiveness of our proposed methods. The implementation is handled in Matlab R2013a on a PC with 2.39 GHz Core 2 Duo processor and with 2 GB of memory. The proposed fusion method is evaluated on different datasets each includes pre-registered CT and MRI images of the same person and the same part of the body. Furthermore, obtained results are compared quantitatively and qualitatively with other existing methods of the literature according to several performance measures.

### 4.1 Evaluation Criterion

Visual perception is most of time subjective when providing instinctive comparisons of the fused images due to eyesight level and mental state. As a consequence, several evaluation metrics should be applied in order to provide an objective assessment. These criterions are of two types; metrics based on single image and the others integrating both source and fused images.

**Entropy (En).** Entropy measures the amount of information available in both source and fused images each apart. The larger is the entropy of the fused image denotes the presence of more abundant information. It is defined as follows:

$$En = -\sum_{i=0}^{l-1} p(i)log_2 p(i) \tag{7}$$

Where $p(i)$ indicates the probability of pixels gray level with the range $[0,\ldots,l-1]$.

**Standard deviation (STD).** STD reflects the contrast of a single image. An image with high standard deviation will have high contrast. The degree of deviation between pixels gray level of an image $I(i,j)$ whose size is $M \times N$ and the average is expressed by:

$$STD = \sqrt{\sum_{i=1}^{M}\sum_{j=1}^{N}\frac{\left[I(i,j) - \left((1/(M \times N)\sum_{i=1}^{M}\sum_{j=1}^{N}I(i,j)\right)\right]^2}{M \times N}} \tag{8}$$

**Spatial frequency (SF).** Spatial frequency (SF) [16] reflects the level of clarity and returns the whole activity of an image. Hence, the larger is the SF the higher is the resolution. It is calculated trough row and column frequency and defined as:

$$SF = \sqrt{RF^2 + CF^2} \tag{9}$$

Where RF is row frequency and CF is column frequency both defined by Eqs. 10 and 11 where a and b denotes the image size and $I(i,j)$ gives the gray level of the fused image.

$$RF = \sqrt{\frac{1}{a(b-1)}\sum_{i=1}^{a}\sum_{j=2}^{b}(I(i,j-1) - I(i,j))^2} \tag{10}$$

$$CF = \sqrt{\frac{1}{(a-1)b}\sum_{i=2}^{a}\sum_{j=1}^{b}(I(i,j) - I(i-1,j))^2} \tag{11}$$

**Structural similarity index (SSIM).** SSIM [29] is a perceptual metric that quantifies image quality degradation. It expresses the similarity between the reference and the fused image and it values is in $[-1, 1]$. So that large value means similarity between source and fused images and the value 1 indicates the identical between two images. It is defined as:

$$SSIM(F,I) = \frac{((2\mu_F\mu_I + C_1) \times (2\sigma_{FI} + C_2))}{((\mu_F^2 + \mu_I^2 + C_1) \times (\sigma_F^2 + \sigma_I^2 + C_2))} \tag{12}$$

Where F is the fused image, I is the input image, $\mu_F$ and $\mu_I$ are respectively the mean intensity of image F and I, $\sigma_F$ and $\sigma_I$ denotes the variance of image F and I, $\sigma_{FI}$ calculates the covariance of F and I and finally, $C_1$ and $C_2$ are constants.

**Peak signal to noise ratio (PSNR).** PSNR is given in dB value for quality judgment and it reflects the level of noise restraint. Better fused image quality is related to the higher value of PSNR which means little difference between input and fused image and less distortion. It is expressed by:

$$PSNR = 10 \times \log_{10}(255^2/MSE) \tag{13}$$

**Mutual information (MI).** MI indicates how much information that input image brings to the fused image. Its value increases with increasing of details and texture information in the fused result. Given two input images $X_A, X_B$ and a fused image $X_F$ It is defined as [16]:

$$MI = I(X_A; X_F) + I(X_B; X_F) \tag{14}$$

Where,

$$I(X_R; X_F) = \sum_{u=1}^{L} \sum_{v=1}^{L} h_{R,F}(u, v) log_2 \frac{h_{R,F}(u, v)}{h_R(u)h_F(v)} \tag{15}$$

R denotes a reference image and F a fused image, where $h_{R,F}(u, v)$ is the joint gray level histogram of $X_R$ and $X_F$. $h_R(u), h_F(v)$ are the normalized gray level histogram of $X_R$ and $X_F$ respectively.

**Image quality index (IQI).** IQI reflects the quality of the fused image. Its dynamic range is $[-, 1]$ and IQI is higher closer to unit signifies the better quality of the fused result. IQI is defined as:

$$IQI = \left(\frac{\sigma_{FR}}{\sigma_F \sigma_R}\right) \cdot \left(\frac{2\mu_F \mu_R}{\mu_F^2 + \mu_R^2}\right) \cdot \left(\frac{2\sigma_F \sigma_R}{\sigma_F^2 + \sigma_R^2}\right) \tag{16}$$

Where $\mu_F, \mu_R$ are the means and $\sigma_F, \sigma_R$ are the variances of fused and source images, respectively. Since two source images A and B are contributing in the fusion process, so the total IQI value is given by the mean:

$$IQI = \frac{IQI(A, F) + IQI(B, F)}{2} \tag{17}$$

## 4.2  Results and Discussion

Experiments are carried out on different datasets including CT and MR images in order to compare the proposed approach with several existing methods. It is obvious that CT images discriminate soft tissues information and show bone structures where the MR images provide the soft tissue information and lacks in boundary information. In the following, experiments are conducted on different datasets and results will be discussed quantitatively and qualitatively based on performance metrics described above and also on visual perception.

**Experiment 1.** Visual and quantitative results of three methods dealing with CT/MR image fusion were compared with the proposed method. Iterative Neuro-Fuzzy Approach (INFA), Discrete Wavelet Transform (DWT) based approach and Lifting Wavelet Transform combined with Neuro-Fuzzy Approach (LWT-NFA) [20] are analyzed and compared subjectively and objectively based on EN and SSIM. Performance results are listed in Tables 1 and 2. Comparative analysis is carried on six different pairs of pre-registered CT (Fig. 6. A1-F1) and MR (Fig. 6. A2-F2) images

**Table 1.** Entropy performance of different approaches.

| | Methods | | | |
|---|---|---|---|---|
| | INFA | DWT | LWT-NFA | Proposed method |
| Dataset 1 | 6.4787 | 6.1688 | 6.35ll | **6.8144** |
| Dataset 2 | 6.4717 | 6.1081 | 6.3675 | **6.7004** |
| Dataset 3 | 6.3794 | 6.0444 | 6.2601 | **6.7271** |
| Dataset 4 | 6.4605 | 6.1397 | 6.3631 | **6.8463** |
| Dataset 5 | 6.5104 | 6.2040 | 6.4176 | **6.8506** |
| Dataset 6 | 6.4275 | 6.0945 | 6.2936 | **6.7803** |

**Table 2.** SSIM performance of different approaches.

| | Methods | | | |
|---|---|---|---|---|
| | INFA | DWT | LWT-NFA | Proposed method |
| Dataset 1 | 0.6362 | 0.2250 | 0.5031 | **0.7784** |
| Dataset 2 | 0.6241 | 0.2145 | 0.4658 | **0.7261** |
| Dataset 3 | 0.6377 | 0.2124 | 0.4850 | **0.7382** |
| Dataset 4 | 0.6574 | 0.2172 | 0.5113 | **0.7227** |
| Dataset 5 | 0.6338 | 0.2206 | 0.4990 | **0.7448** |
| Dataset 6 | 0.6492 | 0.2193 | 0.5122 | **0.7898** |

(256 × 256). Their resultant fusion images are shown in Fig. 6. From the visual analysis of the fused results, it can be observed that our method preserve successfully both soft tissue information provided by MR images and bony structures given by CT images with better resolution compared with the aforementioned methods.

As mentioned above, besides the visual comparison, an assessment of quantitative results based on evaluation criterion (EN and SSIM) demonstrates the outcomes of the proposed fusion method. Table 1 shows the entropy results where it can be concluded that the proposed method gives the highest performance than others which means that more information lies in fused image given by our algorithm. Table 2 exposes the highest values of SSIM produced by the proposed method over different methods. It reveals that our algorithm produces less quality degradation of the resultant image which means the better similarity between source and fused image.

**Experiment 2.** To further evaluate and compare the performance of the proposed methods with the adjacent literature, we propose to process another pre-registered image set (Fig. 7) already applied by several methods. The fusion results are compared with neuro-fuzzy based fusion method in the non-subsampled contourlet domain [18] (NSCT-NF), neuro-fuzzy approach [25] (INF), non-subsampled shearlet transform and spiking neural network [16] (NSST-NN), pulse coupled neural network in the non-subsampled shearlet domain [26] and finally, shearlet transform based fusion approach [13]. Objective evaluation of different results is tabulated in Table 3. While visual results are demonstrated in Fig. 7.

On the basis of visual results given by different scheme and illustrated in Fig. 7, it can be recognized that the proposed fusion algorithm produces fused images with competitive quality and containing both soft tissue and dense tissue information derived from source images. Further, edges information is recuperated in resultant image with good contrast. Additionally, objective evaluation performance listed in Table 3 shows greatest entropy produced by the proposed scheme means that more information is preserved. The standard deviation value is competitive with other methods reflecting a good contrast compared to others as assessed by visual perception. Mutual information and spatial frequency values are not the better compared to the rest of scheme due to the training of input data but still higher than neuro-fuzzy in the non-subsampled contourlet domain. IQI provided by our algorithm is greatest compared to other schemes depicting the better similarity between reference and fused

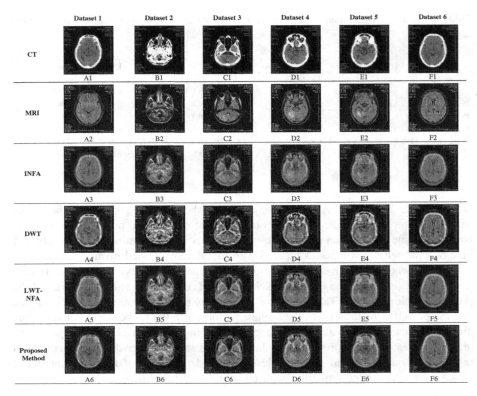

**Fig. 6.** Comparative visual results of different methods applied to input CT (A1 - F1) and MR (A2 - F2) images. Rest of rows illustrates fusion results provided by INFA method (A3 - F3), DWT method (A4 - F4), LWT-NFA method (A5 - F5) and the proposed method (A6 - F6).

**Table 3.** Comparative performance of different methods for the dataset shown in Fig. 7.

|  | EN | STD | MI | SF | IQI | PSNR | SSIM |
|---|---|---|---|---|---|---|---|
| NSCT-NF [18] | 6.7918 | 64.6989 | – | 7.2512 | – | – | – |
| NFA [25] | 4.4894 | – | – | 16.9926 | 0.3182 | 11.4129 | – |
| NSST-NN [16] | 6.8352 | 62.1700 | 4.1550 |  | – | – | – |
| NSST-MAX- SF - PCNN [26] | 6.7801 | 60.0200 | 3.7930 |  | – | – | – |
| ST [13] | 6.1851 | 45.0704 | – |  | – | – | 0.6881 |
| Proposed scheme | **7.1030** | **64.7186** | 3.5718 | 7.2515 | **0.5227** | **20.5818** | **0.8024** |

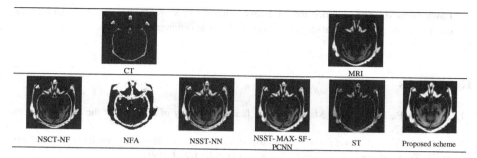

**Fig. 7.** Fusion results of different methods applied to input CT and MR images. Rest of rows illustrates fusion results provided by NSCT-NF, NFA, NSST-NN-PCNN, ST and the proposed scheme.

images. Finally, PSNR and SSIM are comparatively better. Thus, the new method decreases noise than others and the corresponding fusion results are similar to references with less distortion. It can also be concluded that pseudo-Gibbs phenomenon is suppressed through shift invariant shearlet transform.

Time cost is also paid attention in this work. It has been summarized that NSST is lower time consuming than NSCT [16]. Moreover, fusion rules based on PCNN are time consuming due to the learning process compared to the average or maximum methods but still closer to neuro-fuzzy fusion rule.

## 5  Conclusion and Perspectives

Equipped with a rich mathematical structure, shearlet transform is an MGA tool that possesses anisotropy, directionality and shift invariance. In this work, we have exposed a multimodal medical image fusion method based on non-subsampled shearlet transform and neuro-fuzzy. Thus, low frequency sub-bands are fused by maximization of absolute value while high frequency fusion rule is based on Neuro-Fuzzy Inference System. Experiments carried on different CT and MR pre-registered datasets reveals the effectiveness of the proposed method. Based on visual perception, we can notice that the fused images produced by the proposed scheme are rich of information details that belong both to soft tissues and bones with good contrast. Objective evaluation demonstrates that the fusion results provided by the proposed method contain more details and less distortion and noise. Subsequently, the main advantage of the shift invariant shearlet transform over standard shearlet is covered which is the elimination of the pseudo-Gibbs phenomenon.

Additional outcomes are attempted in future in order to further optimize and enhance the performance of our method. Fusion rules for low and high frequency will be addressed with more attention and hybrid intelligence will be paid more consideration. Future works will investigate the deep learning in medical image fusion where different modalities will be integrated in the experimental protocol.

**Acknowledgment.** The authors would like to thank Dr S. Rajkuma, the VIT University, Vellore-India, for providing image datasets of patients at different modalities.

# References

1. James, A.P., Belur, V.D.: Medical image fusion: a survey of the state of the art. Inf. Fusion **19**, 4–19 (2014)
2. Mallat, S.G.: A theory for multiresolution signal decomposition: the wavelet representation. IEEE Trans. Pattern Anal. Mach. Intell. **11**(7), 674–693 (1989)
3. Wang, A., Sun, H., Guan, Y.: The application of wavelet transform to multi-modality medical image fusion. In: IEEE International Conference on Networking, Sensing and Control, Ft. Lauderdale, FL, pp. 270–274 (2006)
4. Yang, L., Guo, B.L., Ni, W.: Multimodality medical image fusion based on multiscale geometric analysis of contourlet transform. Neurocomputing **72**(1–3), 203–211 (2008)
5. Mankar, R., Daimiwal, N.: Multimodal medical image fusion under non-subsampled contourlet transform domain. In: International Conference on Communications and Signal Processing (ICCSP), 2015, Melmaruvathur, pp. 0592–0596 (2015)
6. Jothi, C., Elvina, N., Vetrivelan, P.: Medical image fusion using ridgelet transform. In: International Conference on Innovations in Intelligent Instrumentation, Optimization and Signal Processing, pp. 21–25 (2013)
7. Lu, H.M., Nakashima, S., Li, Y.J., Zhang, L.F., Yang, S.Y., Seiichi, S.: An improved method for CT/MRI image fusion on bandelets transform domain. Appl. Mech. Mater. **103**, 700–704 (2012)
8. Ali, F.E., El-Dokany, I.M., Saad, A.A., Abd El-Samie, F.E.: A curvelet transform approach for the fusion of MR and CT images. J. Modern Optics **57**(4), 273–286 (2010)
9. Kutyniok, G., Labate, D.: Introduction to shearlets. In: Kutyniok, G., Labate, D. (eds.) Shearlets: Multiscale Analysis for Multivariate Data, Birkhäuser, Boston (2012)
10. Do, M.N., Vetterli, M.: The contourlet transform: an efficient directional multiresolution image representation. IEEE Trans. Image Process. **14**(12), 2091–2106 (2005)
11. Easley, G., Labate, D., Lim, W.Q.: Sparse directional image representations using the discrete shearlet transform. Appl. Comput. Harmonic Anal. **25**(1), 25–46 (2008)
12. Guo, K., Labate, D., Lim, W.Q.: Edge analysis and identification using the continuous shearlet transform. Appl. Comput. Harmonic Anal. **27**(1), 24–46 (2009)
13. Miao, Q.G., Shi, C., Xu, P.F., Yang, M., Shi, Y.B.: A novel algorithm of image fusion using shearlets. Optics Commun. **284**(6), 1540–1547 (2011)
14. Deng, C., Wang, S., Chen, X.: Remote sensing images fusion algorithm base on shearlet transform. In: Proceeding of International Conference on Environmental Science and Information Application Technology, pp. 451–454. ACM, Wu Han, China, (2009)
15. Wang, L., Li, B., Tian, L.F.: EGGDD: an explicit dependency model for multi-modal medical image fusion in shift-invariant shearlet transform domain. Inf. Fusion **19**, 29–37 (2014)
16. Singh, S., Gupta, D., Anand, R.S., Kumar, V.: Non-subsampled shearlet based CT and MR medical image fusion using biologically inspired spiking neural network. Biomed. Signal Process. Control **18**, 91–101 (2015)
17. Teng, J., Wang, S., Zhang, J., Wang, X.: Neuro-fuzzy logic based fusion algorithm of medical images. In: 3rd International Congress on Image and Signal Processing (CISP), 2010, Yantai, pp. 1552–1556 (2010)

18. Das, S., Kundu, M.K.: A neuro-fuzzy approach for medical image fusion. IEEE Trans. Biomed. Eng. **60**(12), 3347–3353 (2013)
19. Kavitha, C.T., Chellamuthu, C.: Multimodal medical image fusion based on Integer Wavelet Transform and Neuro-Fuzzy. In: International Conference on Signal and Image Processing (ICSIP), 2010, pp. 296–300 (2010)
20. Rajkumar, S., Bardhan, P., Akkireddy, S.K., Munshi, C.: CT and MRI image fusion based on Wavelet Transform and Neuro-Fuzzy concepts with quantitative analysis. In: International Conference on Electronics and Communication Systems (ICECS), 2014, Coimbatore, pp. 1–6 (2014)
21. Guo, K., Lim, W., Labate, D., Weiss, G., Wilson, E.: Wavelets with composite dilation s and their MRA properties. Appl. Comput. Harmonic Anal. **20**(2), 231–249 (2006)
22. Liu, S., Shi, M., Zhu, Z., Zhao, J.: Image fusion based on complex-shearlet domain with guided filtering. Multidimension. Syst. Signal Process. **20**(2), 1–18 (2015)
23. Kong, W.W., Liu, J.P.: Technique for image fusion based on non-subsampled shearlet transform and improved pulse-coupled neural network. Optical Eng. **52**(1), 017001/1–12 (2013)
24. Balasubramaniam, P., Ananthi, V.P.: Image fusion using intuitionistic fuzzy sets. Inf. Fusion **20**, 21–30 (2014)
25. Rao, D.S., Seetha, M., Hazarath, M.: Iterative image fusion using neuro fuzzy logic and applications. In: International Conference on Machine Vision and Image Processing (MVIP), 2012, Taipei, pp. 121–124 (2012)
26. Geng, P., Wang, Z., Zhang, Z., Xiao, Z.: Image fusion by pulse couple neural network with shearlet. Optical Eng. **51**, 067005 (2012)
27. Zagrouba, E., Barhoumi, W.: Semiautomatic detection of tumoral zone. Image Anal. Stereology **21**(1), 13–18 (2002)
28. Shearlet webpage. www.shearlab.org
29. SSIM. http://www.ece.uwaterloo.ca/~z70wang/research/ssim/

# Author Index

Printed in the United States
By Bookmasters